Bio-Materials and Prototyping Applications in Medicine

Paulo Jorge Bártolo • Bopaya Bidanda
Editors

Bio-Materials and Prototyping Applications in Medicine

Second Edition

Editors
Paulo Jorge Bártolo
Centre for Rapid and Sustainable Product
Polytechnic Institute of Leiria
Leiria, Portugal

Bopaya Bidanda
Department of Industrial Engineering
University of Pittsburgh
Pittsburgh, PA, USA

ISBN 978-3-030-35878-5 ISBN 978-3-030-35876-1 (eBook)
https://doi.org/10.1007/978-3-030-35876-1

This Springer imprint is published by the registered company Springer Nature Switzerland AG
The registered company address is: Gewerbestrasse 11, 6330 Cham, Switzerland

This book is dedicated to our parents
Lucilia Pinto Dias and Antonio Dias (late)
Neena and Monapa Bidanda (late)
Claudina Coelho da Rocha and Arlindo Terreira Galha (late)
Maria Alice and Francisco Bartolo
And our families
Helena and Pedro
Louella, Maya, and Rahul
For their constant support throughout this project

Finally, we would like to extend a special acknowledgment to
Dr. Fengyuan Liu
Without her untiring efforts this book would never have been completed

Preface

We are especially pleased to present the second edition of our edited book in an area that is quickly emerging as one of the most active research areas that integrates both engineering and medicine. Research in this area is growing by leaps and bounds. Preliminary research results show significant potential in effecting major breakthroughs ranging from a reduction in the number of corrective surgeries needed to the "scientific miracle" of generating tissue growth. Billions of dollars/euros/pounds have been invested in tissue engineering over the past decade – a large and significant component of this is in the area of biomaterials and prototyping applications in medicine. As a result, we have made significant changes to this edition as you will see from our summary of chapters below.

Chapter 1 by Desai et al. discusses the different types of biomaterials used for medical applications. Metallic-, ceramic-, and nanomaterial-based biomaterials are classified and described based on their physical and biocompatibility properties. Various applications of these biomaterials and the state of the art in biomaterial research are discussed. A new class of biodegradable magnesium alloy with potential to replace and augment existing implant materials is elaborated.

In Chap. 2, Salil Desai and M. Ravi Shankar review the use of synthetic polymers, carbon-based composites, and carbon nanotube-based biomaterials and discuss the implementation of hydrogels for drug delivery carriers and tissue engineering applications. Furthermore, the enhancement of mechanical properties with nano-filled polymers for dental restorations and structural applications is elaborated. The efficacy of bio-functional coatings with elastomeric polymers loaded with antiproliferation drugs for cardiovascular devices is presented.

Tomaraei et al. discuss the current status and existing opportunities for leveraging silk and silk-derived materials for biomedical applications in Chap. 3. The process-structure-property relationship for natural silk and regenerated silk materials in the form of films, fibers, spheres, hydrogels, sponges, and tubes is presented. They also explore a number of silk-derived materials that either include a second phase in addition to silk such as in the case of composites or that chemically transform silk into another material such as in the case of carbonization.

Four-dimensional (4D) bioprinting has emerged in recent years, where "time" is integrated as the fourth dimension, enabling the printed product with dynamic behaviors and constantly shape-changing features by applying an external stimulus, e.g., temperature, water, and light. In Chap. 4, Moataz and Youngjae present the use of shape memory alloy, especially nitinol material, in a wide range of medical applications, such as endovascular devices. The mechanical and biological properties of nitinol, various endovascular devices, and commercially available nitinol endovascular devices, including guidewires, stents, percutaneous heart valves, occluders, and filters, are reviewed and discussed.

In Chap. 5, Mohammed et al. discuss the use of industrial optical scanning methods to capture the surface topology from a volunteer's facial anatomy and demonstrate the potential of the 3D design and multi-model/multi-material printing, augmented with the use of optical surface scanning, to produce realistic prosthetic models of both the ear and nose. The novel design techniques enable tailoring the skin pigmentation of the prosthesis to a variety of skin tones and mimicking the mechanical properties of the original anatomy.

Geng and Bidanda provide an overview of the entire additive manufacturing (AM) processes adopted by medical community and review the current state-of-the-art medical applications of AM techniques in Chap. 6. The comparison of traditional manufacturing processes and AM processes and the limitations and future trends of AM in medical applications are presented.

In Chap. 7, Mishbak et al. review and discuss the recent advances in the design of naturally derived photocrosslinkable hydrogels for cartilage tissue engineering. A brief description of the structure and composition of cartilage tissue is provided, and an overview of hydrogel properties, synthesis routes, and strategies for hydrogel biofunctionalization is described. They also discuss the recent applications of common photocrosslinkable natural hydrogels in cartilage tissue, along with the manufacturing technologies to design hierarchical cartilage-like tissue constructs.

In Chap. 8, Alqahtani et al. present a detailed state of the art of exoskeletons for lower limb applications by providing a classification based on the design purposes, including rehabilitation, augmentation, and locomotion assistance, and presenting a large number of different types of exoskeletons. The major research challenges and opportunities are also presented.

In Chap. 9, Daskalakis et al. discuss the key characteristics of bioglass materials and present the main technologies being used to fabricate scaffolds incorporating bioglass materials due to the high bioactivity of bioactive glasses and the potential stimulation of osteogenesis and angiogenesis for bone tissue engineering. Scaffold design requirements, including structural properties, porosity, surface topography, and wettability, and the traditional and AM scaffold fabrication techniques are introduced and discussed.

The production of this book has been a most enjoyable experience. We thank the authors for their valuable and timely contributions to this volume. We also would like to thank Prof. Martin Schröder, Dean of the Faculty of Science and Engineering; Prof. Alice Larkin, Head of the School of Engineering; Prof. Tim Stallard, Head of the Department of Mechanical, Aerospace, and Civil Engineering, University of

Manchester; and Dean Jimmy Martin, US Steel Dean of Engineering, University of Pittsburgh, for their support of our academic endeavors.

We would especially like to acknowledge the unlimited patience and constant support of Brinda Megasyamalan of Springer. Finally, we would like to acknowledge the multiple contributions of Dr. Fengyuan Liu. Without her untiring efforts, abundance of patience at our tardiness, and outstanding project management skills, this edition would never have been completed. Thank you Fengyuan!

Leiria, Portugal
Pittsburgh, PA, USA

Paulo Jorge Bártolo
Bopaya Bidanda

Contents

Chapter 1
Emerging Trends in the Applications of Metallic and Ceramic Biomaterials

Salil Desai, Bopaya Bidanda, and Paulo Jorge Bártolo

1.1 Preface

Materials that interface with biological entities and are used to create prosthesis and medical devices and replace natural body tissue are broadly called biomaterials. An expert definition of biomaterial is [1, 2] (Williams, 1987, Ratner et al., 2004):

> A biomaterial is a nonviable material used in a medical device, intended to interact with biological systems.

Though the initial definition of biomaterial was restricted to medical devices, biomaterials in the present times encompass both synthetic and natural materials that promote human health. A distinctive difference between a biomaterial over other materials is its benign coexistence with a biological system with which it interfaces. This phenomenon is called as biocompatibility and is defined as [1] (Williams, 1987):

> Biocompatibility is the ability of a material to perform with an appropriate host response in a specific application.

The biocompatibility of materials is of considerable interest because implants and tissue interfacing devices can corrode in an in vivo environment [3]. The corro-

S. Desai (✉)
Department of Industrial & Systems Engineering, North Carolina A&T State University, Greensboro, NC, USA
e-mail: sdesai@ncat.edu

B. Bidanda
Department of Industrial Engineering, University of Pittsburgh, Pittsburgh, PA, USA
e-mail: bidanda@pitt.edu

P. J. Bártolo
Centre for Rapid and Sustainable Product Polytechnic Institute of Leiria, Leiria, Portugal
e-mail: paulojorge.dasilvabartolo@manchester.ac.uk

P. J. Bártolo, B. Bidanda (eds.), *Bio-Materials and Prototyping Applications in Medicine*, https://doi.org/10.1007/978-3-030-35876-1_1

sion of the implant can lead to loss of load bearing strength and consequent degradation into toxic products within the tissue.

Biomaterial applications span from prostheses (e.g., hip implants and artificial heart valves), tissue regeneration, medical devices, to drug delivery. A biomaterial is specifically chosen based on its compatibility with the host tissue and structural integrity over its designed life [4–8]. In order to identify each material type based on their properties and application intent, it is important that they be systematically classified. Biomaterials can be broadly categorized under the four categories, namely:

- Metals
- Ceramics
- Polymers
- Composites

1.2 Metals

This class of material is known for their high stiffness, ductility, wear resistance, and thermal and electrical conductivity. Metals and their alloys are commonly used in implants, medical device manufacture, and related accessories. Due to their mechanical reliability, metallic biomaterials are difficult to be replaced by ceramic and polymer substitutes [9]. One of the advantages of using metals as biomaterials is their availability and relative ease of processing from raw ore to finished products. The material properties of metals have been studied in the context of biocompatibility, surface interaction, and structural integrity [3, 10–13]. Moreover, customized properties including flexibility, high strength, and abrasion resistance can be developed by alloying constituent elements of different metals. Metallic biomaterials are classified as inert because they illicit minimal tissue response. Given their higher fatigue strength and chemical resistance to corrosion, they are used in load-bearing applications. This section describes the different types of metals and their alloys that are commonly used.

1.2.1 Titanium

1.2.1.1 Description

Titanium has one of the highest strength-to-weight ratio and corrosion resistance of metals [14, 15]. It has a lustrous metallic-white color and exhibits high hardness. In its pure form, titanium is ductile [16] and is often alloyed with other elements for enhanced toughness. Titanium is extracted from rutile (TiO_2), a mineral deposit, and is processed in multiple steps [17, 18] to obtain the finished material. Due to its noncorrosive properties, titanium has excellent biocompatibility. The material passivates itself in vivo by the formation of an adhesive oxide layer [19, 20]. Titanium also displays a unique property of osseointegration where it connects both structurally and functionally with the underlying bone [21]. It is commonly used in total joint replacements [22], dental implants [23], internal and external fixators, artificial heart valves, spinal fusion, and medical devices. However, due to the high processing cost, titanium is expensive.

1.2.1.2 Applications

Due to its high strength, low weight, and noncorrosive properties, titanium and its alloys are used in a wide range of medical applications. Titanium is a major material used in the skeletal system for joint replacement such as hip ball and sockets and in internal fixators such as plates and screws. A titanium implant has high fracture toughness and enhanced fatigue properties over competing metals. These load-bearing implants can stay in place for 15 to 20 years thereby, improving the quality of human life. Figure 1.1 shows the acetabular shell (socket portion) and the femoral stem. The socket is made of metal shell with a medical grade plastic liner which acts like a bearing. The femoral stem is made of metal such as titanium alloy.

Figure 1.2 shows one of the femoral stems being cemented to the bone using an epoxy. In the other design, the femoral stem has a fine mesh of holes on the surface that promotes tissue growth and subsequent attachment of the prosthesis to the bone.

Titanium is also used for bone-fracture fixation in spinal fusion devices, pins, bone plates, and screws. Due to its nonmagnetic properties, it does not pose any threat to patients with implants during magnetic resonance imaging and exposure to electronic equipment. Titanium is also used for a wide range of surgical instruments. It does not corrode or lose surface properties with repeated sterilization and its light weight reduces surgeon fatigue during repetitive operations. Titanium is used in craniofacial and maxillofacial treatments to replace facial features of patients.

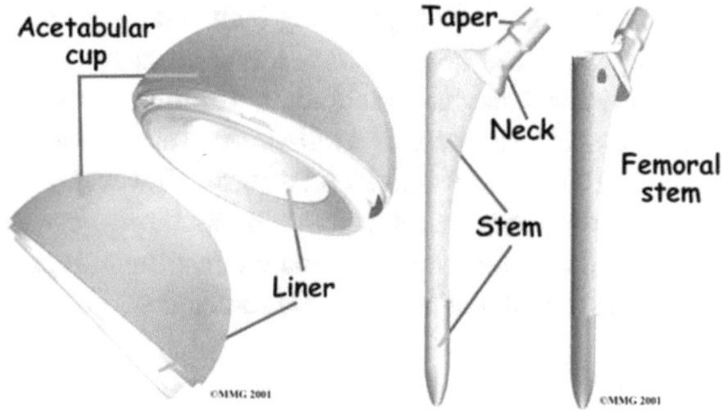

Fig. 1.1 Acetabular shell (socket-left) and femoral stem (right) implants. (Courtesy: DePuy Orthopaedics, Inc. [24])

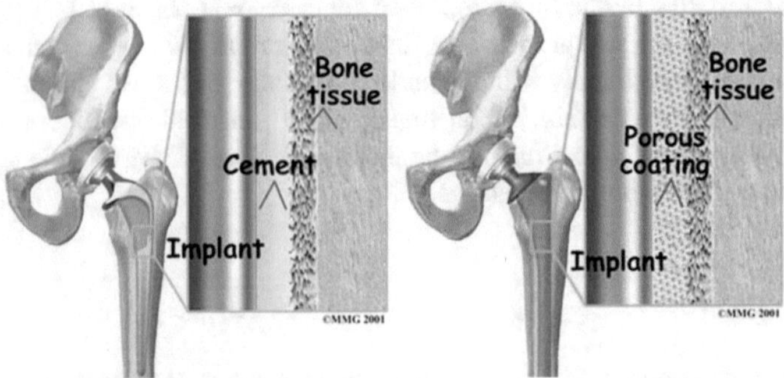

Fig. 1.2 Cemented (left) and uncemented (right) designs of femoral stem implants. (Courtesy: DePuy Orthopaedics, Inc. [24])

Another prominent application of titanium alloy is in dental implants for tooth fixation. After the osseointegration of the implant with the bone, an abutment is inserted into the implant. The abutment provides a seat for the crown which replaces the natural tooth. Tooth replacement using titanium implants is more effective than the use of traditional root canal and bridge constructs.

1.2.2 Stainless Steel

1.2.2.1 Description

Stainless steel is a versatile class of material that has high strength and resistance to oxidation. Typically, stainless steels have a minimum of 12% chromium content that forms a thin oxide film which resists oxidation. The addition of nickel and molybdenum further enhances the corrosion and pitting resistance. As compared with titanium, it is easy to machine and thus commonly used for surgical instruments, bone screws, stents, and other medical equipment. Of the numerous grades of stainless steels, the 300 series is used in medical applications. Typical medical grades include the 304 and the 316 L. It can be also electropolished for aesthetic appeal. Because of its high strength and chemical inertness to bodily fluids, blood, and enzymes, it is FDA approved as a biomaterial. Moreover, it can be processed using multiple methods including forming, welding, bending, and machining. Medical-grade stainless is available in various stock forms, making it is easy to fabricate the material into its final form. However, it is heavier than titanium which can lead to heavier implants and fatigue during repeated handling of surgical tools. Based on their microstructure and the resulting properties, steels are broadly classified as austenitic, martensitic, and ferritic.

1.2.2.2 Applications

Commercial-grade stainless steel is used to manufacture operating room accessories and dental and surgical instruments which involve superficial contact of the device with the human tissue. Austenitic steels are used for implant fabrication, hypodermic needles, sterilizers, work tables, and autoclave compartments where moderate strength, formability, and corrosion resistance is desired [25]. This class of stainless steel is nonmagnetic, can be cold hardened, and possesses higher corrosion resistance than other types. 316 L is the most common stainless steel used in medical industry. The "L" within the 316 L designation stands for low carbon steel. As compared with the 0.08% carbon content in regular 304 or 316 steel, the 316 L contains 0.03% carbon. The lower carbon content reduces carbide precipitation, thereby minimizing in vivo corrosion. Higher percentages of nickel (~12%) are added to stabilize the austenitic phase of steel. Other alloying elements include chromium, molybdenum, manganese, silicon, sulfur, phosphorous, and nitrogen. Type 316 L austenitic steels can be heat treated for a wide range of mechanical properties. Because they corrode under highly stressed and oxygen-depleted [26] environments, they are generally used for temporary implant devices. BioDur® 108 is a nickel-free austenitic stainless steel alloy with high nitrogen content [27]. It has higher tensile and fatigue strength as compared with nickel-containing alloys such as 316 L. It is a nonmagnetic alloy that can be fabricated by forging or machining. Due to its high strength and corrosion resistance, it is used in bone plates, spinal fixation, screws, hip and knee components, and medical devices.

Figure 1.3 (left) shows a stainless screw and washer for soft tissue fixation to the bone. Figure 3 (right) shows a low-profile, wide staple-based fixation system that provides better load distribution. The unique staple design enhances uninterrupted vascular flow to the underlying tissue [28].

Martensitic stainless steels contain iron, chromium, and carbon alloys with other additives including niobium, silicon, tungsten, and vanadium. The properties (hardness, toughness) of these stainless steels can be altered based on heat treatment conditions. These types of steels possess lower corrosion resistance over austenitic steels. They are used for dental and surgical instruments such as chisels, scalpels, pliers, forceps, among others. Figure 1.4 shows instruments made from surgical-grade stainless steel.

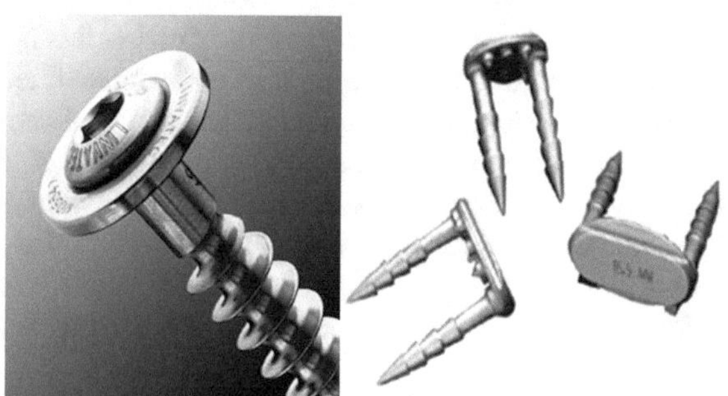

Fig. 1.3 Stainless steel screw and washer (left) for soft tissue fixation, staple fixation system to attach tissue to bone (right). (Courtesy: ConMed Linvatec Corporation [28])

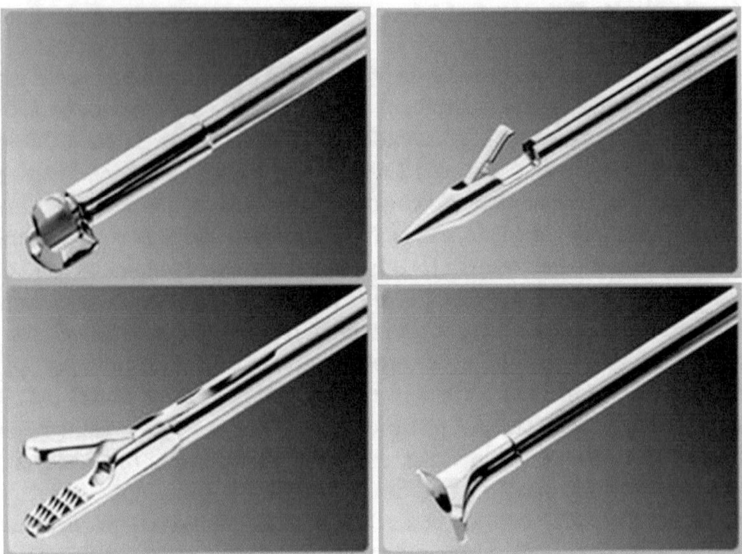

Fig. 1.4 Surgical-grade stainless steel rotary punch (top left), suture tram (top right), linear alligator grasper (bottom left), rotary scissors (bottom right). (Courtesy: ConMed Linvatec Corporation [29])

1.2.3 Shape Memory Alloys

1.2.3.1 Description

Shape memory alloys (SMA) are materials that retain their original shape after severe deformations when subjected to heat above their transformation temperature. NiTiNOL which stands for Ni-Nickel, Ti-Titanium, and NOL-Naval Ordnance Laboratory is a popular shape memory alloy discovered by Buehler et al. at the US Naval Ordnance Laboratory in the 1960s [30]. Shape memory alloys have two distinct crystallographic phases, namely, austenite and martensite. The martensitic phase is a low temperature stable phase with the absence of stress. The austenite phase is stable at high temperature and displays a stronger body-center cubic structure [31]. SMA are capable of large amounts of bending and torsional deformation and high strain rates (6%–8%) in the martensitic phase [32]. Once deformations are induced in the material in the martensitic phase, it is heated at the phase transformation temperature. The alloy undergoes a crystalline reversible solid-state phase change from martensite to austenite. The above phenomenon is known as one-way shape memory effect. Another unique property of shape memory alloy is pseudo-elasticity wherein the two-way phase transformation occurs at a constant temperature A_F. A_F is defined as the temperature where the austenitic phase is finished forming. A shape memory alloy fully composed of austenite phase is mechanically deformed at constant temperature A_F. The loading transforms the material into a martensitic phase. When the load is released, reverse transformation (martensitic to austenitic) occurs, bringing the material back to its original shape. The properties of shape memory alloys are significantly affected by composition, processing methods, and other factors. Nitinol alloys are processed using various powder metallurgy techniques [33]. Nitinol implants are used as hard tissue implants in orthopedics and dentistry because of its porous structure, good mechanical properties, biocompatibility, and shape memory effect [34, 35]. One of the promising processes for manufacturing hard tissue TiNi alloy implants is HIPing [36]. HIPing stands for Hot Isostatic Pressing wherein different material powders are consolidated using heat and high pressure simultaneously. Using the HIPing process, it is possible to attain implant properties such as controlled porosity and elastic modulus nearer to the natural bone. Figure 1.5 shows SEM micrographs of HIPed samples under different heating temperatures and time durations made from elemental powders of Ti (50%) and Ni (50%) composition [37].

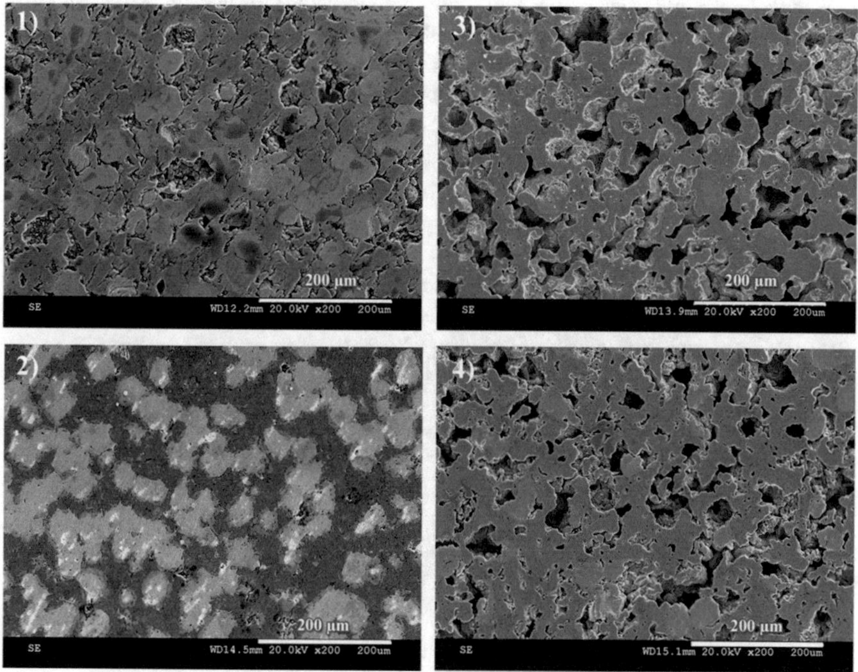

Fig. 1.5 Micrographs show the porous structures of the HIPed specimens. (1) and (2) are for the specimens HIPed at 900 °C for 1 h and (3) and (4) are for those HIPed at 980 °C for 4 h. Specimens (1) and (3) had wax binder in powder preparation, and (2) and (4) had PDDA binder. (Courtesy: Zhigang Xu: Center for Advanced Materials and Smart Structures – NC A&T SU [38])

1.2.3.2 Applications

Shape memory alloys (SMA) possess excellent corrosion resistance, wear, and mechanical properties with good biocompatibility. NiTi-based shape memory alloys are used as biomaterials for in vivo applications including implants and minimally invasive surgeries [39, 40]. This is because NiTi alloys maintain their functional ability without degrading when in contact with living tissue [41]. One of the prominent applications of nitinol alloy is in self-expanding cardiovascular stents. Stents are used in angioplasty procedures to open blocked and weakened blood vessels such as the coronary, iliac, carotid, aorta, and femoral arteries [42]. Stent is a cylindrical scaffold made of shape memory material. Initially, a stent is introduced inside the blood vessel in its pre-compressed martensitic state using a catheter. After reaching the body temperature, the stent expands to open the blood vessel and increase blood flow. Figure 1.6 shows straight and tapered configurations of nitinol stents manufactured by Abbott Vascular Laboratories.

Shape memory alloys are commercially available as alloy combinations of different compositions including copper, zinc, aluminum, cadmium, indium, iron, magnesium, gold, and silver. Copper-based shape memory alloys are used in external biomedical applications which do not need biocompatibility [44]. Applications that exploit the superelasticity and shape-changing properties of shape memory alloys include eyeglass frames, rehabilitation devices, guide wires for introduction of therapeutic and diagnosis devices [45], and snare wires used for removing tonsils and polyps [46, 47]. Other applications of shape memory alloy include vena cava filters that trap blood clots and prevent their flow to other body parts [48, 49]. The device consists of a straight wire that is inserted into the vena cava with a cooled

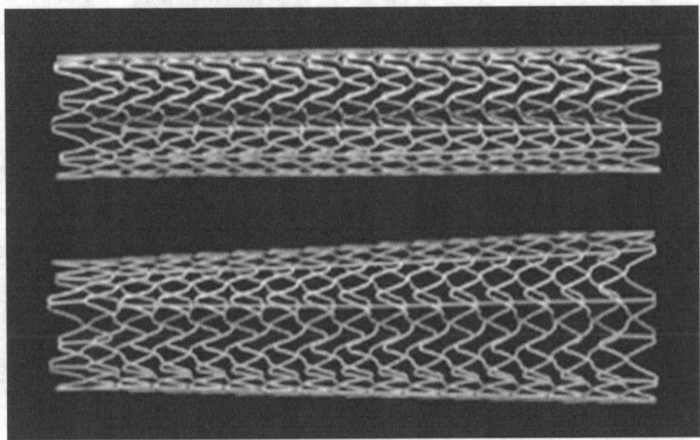

Fig. 1.6 RX Acculink Carotid self-expanding nitinol stents manufactured in straight (top) and tapered (bottom) configurations. (Images courtesy of Abbott Vascular. (c) 2007 Abbott Laboratories. All Rights Reserved [43])

catheter. On reaching body temperature, the wire reverts to an umbrella-shaped filter to trap small blood clots.

SMA are also used for orthodontic archwires to apply a consistent force for correcting misaligned teeth [50, 51]. The dental application utilizes the superelasticity property of shape memory alloys without the need to periodically retighten wires as in conventional stainless steel material.

1.2.4 Noble Metals

Noble metals show a marked reluctance to combine with other elements to form compounds. As such they have excellent resistance to corrosion or oxidation and good candidate for biomaterials. Noble metals such as gold, silver, and platinum are used when there is a need for functionality other than the basic mechanical performance. Typically, they are used in devices requiring specific electrical or mechanical properties.

1.2.4.1 Gold

Gold is an inert metal that has high resistance to bacterial colonization. Gold and its compounds have been historically used in oriental cultures for the treatment of ailments. It has been one of the first materials to be used as an implantable material (dental tooth implant). Due to its high malleability, it is used in restorative dentistry for crowns and permanent bridges. Gold possesses excellent electrical conductivity and biocompatibility and is used in wires for pacemakers and other medical devices.

1.2.4.2 Platinum

Platinum possesses excellent corrosion resistance, biocompatibility, and stable electrical properties. It is used for the manufacture of electrodes in devices such as cardiac pacemakers and electrodes in cochlear (cavity within the inner ear) replacement for the hearing impaired. A typical pacemaker uses platinum–iridium electrodes that send electrical pulses to stabilize the rhythm of the heartbeat. Miniaturized platinum coils are used in endovascular therapy for the treatment of aneurysms. Using a micro-catheter, these platinum coils are inserted within an aneurysm. The flexible platinum coils conform to the shape of the aneurysm and obstruct the flow of blood.

1.2.4.3 Silver

Silver is used in surgical implants and as a sanitizing agent. They are used as studs of earrings to prevent infection of newly pierced ears. Silver compound is used in burn therapy to improve healing and prevent infection of burns. Wound dressing fibers are plated with silver to provide a germicidal and analgesic enclosure on the wound. Silver is also used in urinary bladder catheters and stethoscope diaphragms.

1.2.5 Biodegradable Metals

A new class of materials are being explored in the biomedical implant and device field in the recent years. These include biodegradable metals which provide adequate mechanical strength and bio functionality such as osseointegration within the host site. Biodegradable metals form a desirable choice for the juvenile population wherein the implant needs to be operated at different time points due to the growth of the underlying skeletal frame. Of the prominent metals being considered, magnesium alloys resemble physical and mechanical properties close to that of the natural bone [52]. The implementation of magnesium alloys has shown reduction in stress shielding and promotion of bone growth and remodeling [53]. Magnesium is the fourth abundant mineral in the human body and is an important element in several physiological chemical reactions [54]. In addition, elemental magnesium has shown to be beneficial in tracheal and cardiovascular applications [55, 56]. Magnesium is an ideal candidate for in vivo applications based on its ability to provide the intended function, disintegrate, and get absorbed in the body to be excreted with other waste products. However, pure magnesium has a high corrosion rate when exposed to physiological fluids, resulting in the production of hydrogen gas at the implant site. This problem can be alleviated by alloying pure magnesium and surface modification for its controlled degradation. Biodegradable polymers are used to coat the surface of the magnesium alloy, each providing a barrier layer toward corrosion resistance. Candidate biopolymers for coating include poly(lactic-co-glycolic) acid (PLGA), polycaprolactone (PCL), and polyester urethane urea (PEUU) which can vary the degradation rate of magnesium alloy based on their dissolution properties. Figure 1.7 shows the release of magnesium ions from polymeric coatings with 10 and 20 layers of deposition as compared to an uncoated (bare) magnesium alloy surface. The uncoated (bare) magnesium alloy surface has the highest magnesium ion release due to severe corrosion and pitting of the sample. The PLGA and PEUU coatings offered good protection of the magnesium alloy surface. In addition, coatings with 20 layers had higher corrosion protection as compared with 10 layers. Further, the PEUU coating with 20 layers had the lowest corrosion rate, releasing minimal magnesium ions. On the contrary, the PCL coatings had poor barrier layer protection due to entrapment of fluid media between the coating and magnesium surface.

Magnesium has shown to be beneficial in tracheal applications [58]. Thus, its implementation as a tracheal stent could help restore the thoracic channel in patients suffering from tracheal restenosis. Figure 1.8 shows the time-lapse optical micro-

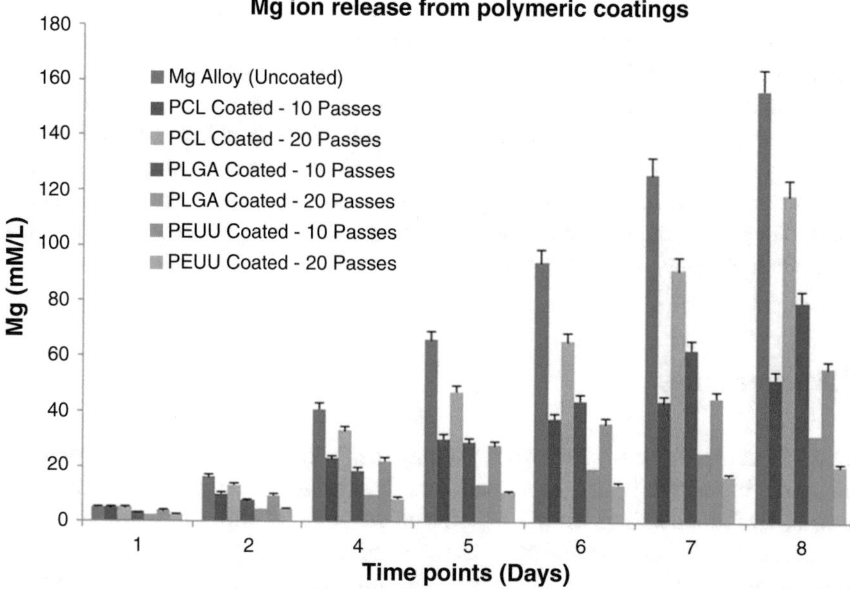

Fig. 1.7 Magnesium ion release from polymer-coated samples with 10 vs. 20 layers compared with uncoated Mg alloy sample in simulated body fluid. (Courtesy of Annals of Biomedical Engineering© 2014 Springer Publishers [57])

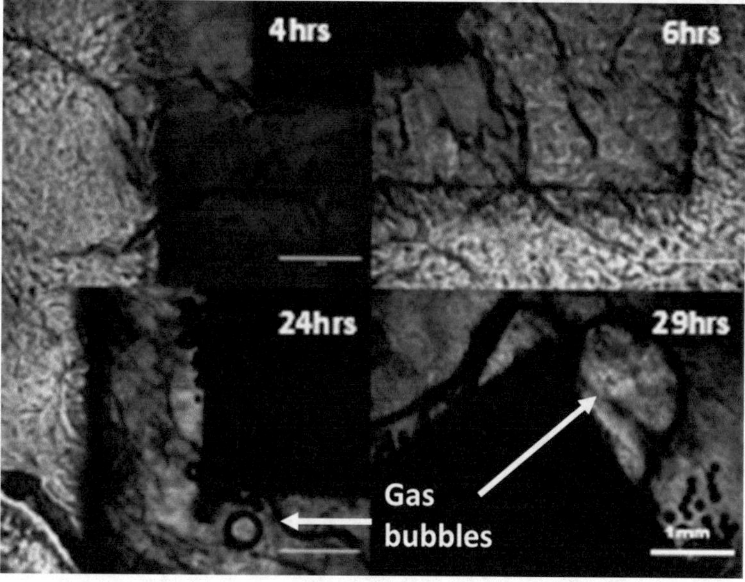

Fig. 1.8 Sample coated with PCL polymer shows release of gas at Mg-NHBE cell interface at different time points. (Courtesy of Annals of Biomedical Engineering© 2014 Springer Publishers [57])

graphs of the PCL-coated samples in culture with fully differentiated normal human bronchial epithelial (NHBE) cells. The PCL-coated magnesium sample resembles the darker region of the image, whereas the NHBE cells and mucus secretion are shown in the background. As can be seen from Fig. 1.8, the interface of PCL-coated magnesium sample does not show the formation of gas bubbles at 4 and 6 hours, respectively. However, the release of gas is seen at 24 hours with progressive bubble formation at 29 hours. Thus, ongoing research is being conducted in the biomedical field to overcome the rapid deterioration of magnesium alloys for its effective usage in implant and device applications.

1.3 Ceramics

Ceramic comes from a Greek word "keramikos" which stands for burnt materials. Ceramic biomaterials are inorganic and nonmetallic elements predominantly formed by ionic bonding. They are produced under high temperature heat treatment process called firing. The use of ceramics as biomaterials has allowed for tailored surfaces that exhibit minimal reaction with host tissues in addition to providing high-bearing properties [59]. Based on their excellent biocompatibility, they are used as implants within bones, joints, and teeth. They are used as coatings in conjunction with metallic core structures for prosthesis. Herein the ceramic provides the hardness and wear resistance, while the metallic core provides toughness and high strength for load-bearing applications [60]. Ceramic structures can be designed with varying porosity for bonding with the natural bone.

1.3.1 Hydroxyapatite (HA)

1.3.1.1 Description

Hydroxyapatite (HA) is a calcium phosphate-based ceramic with high hardness. HA ceramic structures can be developed with a unique bone-like porous structure and are widely used for creating scaffolds in tissue engineering. It can be used for long-term bone replacement due to its slow-decaying properties. Synthetic hydroxyapatite (Ca_{10} $(PO_4)_6(OH)_2$) is an inorganic biomaterial with chemical characteristics similar to hard tissues such as the bone and teeth. Hydroxyapatite ceramics are bioactive, such that they promote hard tissue ingrowth and osseointegration when implanted within the human body [61]. The porous structure of this material can be tailored to suit the interfacial surfaces of the implant. However, they lack mechanical strength for load-bearing applications as stand-alone structural members.

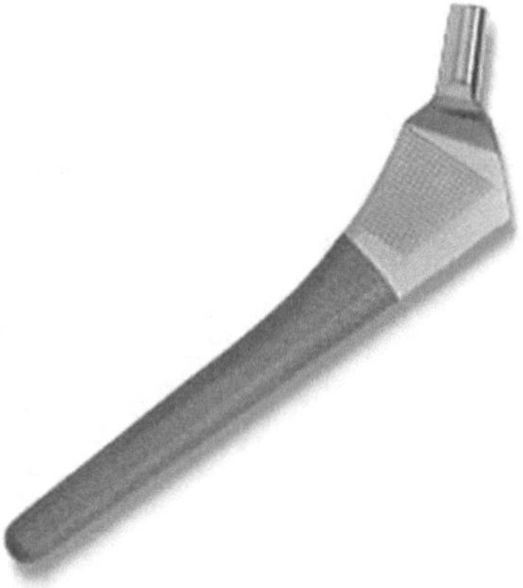

Fig. 1.9 Femoral component coated with hydroxyapatite ceramic coating. (Courtesy: Biomet Orthopedics [62])

1.3.1.2 Applications

The porosity of the hydroxyapatite structure can be controlled similar to the human bone. Thus, it is ideal to be used in implants for artificial tooth, hip, and knee replacements. Typically, most high-bearing implants contain hydroxyapatite coating. The hydroxyapatite coating is applied to the core metallic implant using plasma spray technology. This minimizes the delamination of the hydroxyapatite coating from the metal implant and prolongs the working life of the prosthesis. Figure 1.9 shows the femoral component of the hip implant that has been coated with hydroxyapatite ceramic to promote rapid ingrowth of bone structure. Hydroxyapatite ceramic has also been used in maxillofacial implants as bone filler and as orbital implants within the eye socket.

1.3.2 Zirconia

1.3.2.1 Description

Zirconia (ZrO_2) is a white amorphous powder and dioxide of zirconium. Zirconia is bioinert and thus does not interact with the human body. With an increase in temperature, zirconia changes its monoclinic crystalline state and morphs into tetragonal crystalline and subsequent cubic crystalline. Zirconium oxide ceramic has several advantages over other ceramic materials, due to its transformation-toughening mechanisms, low thermal conductivity, abrasion resistance, desirable biocompatibility, diminished plaque accumulation, and excellent light dynamics [63].

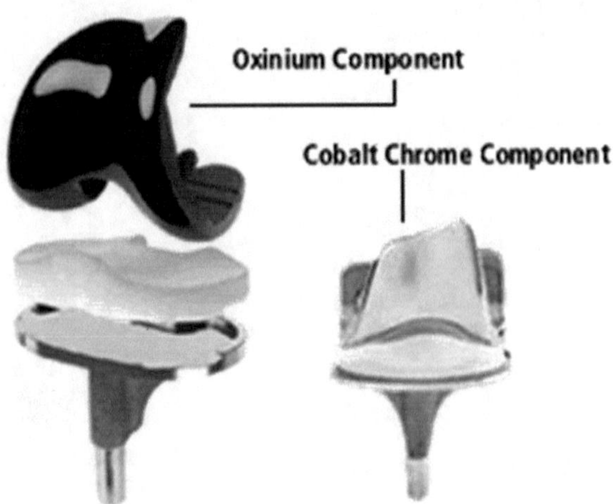

Fig. 1.10 Zirconium oxide (Oxinium™) knee implant. (Courtesy: Smith & Nephew Inc. [68])

1.3.2.2 Applications

Partially stabilized zirconia is commonly used in prosthetic devices because it is stronger and has high resistance to wear. The flexural strength and fracture toughness of zirconia is higher as compared with other ceramics which makes it resistant to masticatory forces when used as crows with exact precision of fit [64, 65]. Also, zirconia implants have shown to accumulate less bacteria in vivo [66] and undergo a lower rate of inflammation-associated processes as compared with titanium [67]. Zirconia has also been used in shoulder reconstruction surgery and as a coating over titanium in dental implants. Yttria-stabilized zirconium oxide implants include knee joints and spinal implants. Figure 1.10 shows zirconium oxide knee implant component manufactured by Smith & Nephew Inc.

Acknowledgments This work was funded by the US National Science Foundation (NSF CMMI: Award 1663128).

References

1. D.F. Williams, Definitions in biomaterials, in *Proceedings of a Consensus Conference of the European Society for Biomaterials, Vol. 4. Chester, England, March 3–5, 1986*, (Elsevier, New York, 1987)
2. B.D. Ratner, S.J. Bryant, Biomaterials: Where we have been and where we are going. Annu. Rev. Biomed. Eng. **6**, 41–75 (2004)
3. D.F. Williams, Orthopedic implants: Fundamental principles and the significance of biocompatibility, in *Biocompatibility of Orthopedic Implants*, ed. by D. F. Williams, vol. 1, (CRC Press, Boca Raton, FL, 1982), pp. 1–50
4. I. Marquetti, S. Desai, Molecular Modeling the adsorption behavior of bone morphogenetic protein – 2 on hydrophobic and hydrophilic substrates. Chem. Phys. Lett. **706**, 285–294 (2018)

5. I. Marquetti, S. Desai, Adsorption behavior of bone morphogenetic protein-2 on a graphite substrate for biomedical applications. Am. J. Eng. Appl. Sci. **11**(2), 1037–1044 (2018). https://doi.org/10.3844/ajeassp.2018.1037.1044
6. A. Aljohani, S. Desai, 3D printing of porous scaffolds for medical applications. Am. J. Eng. Appl. Sci. **11**(3), 1076–1085 (2018)
7. I. Marquetti, S. Desai, Adsorption behavior of bone morphogenic protein (BMP-2) on nanoscale topographies. Proceedings of the ASME NanoEngineering for Medicine and Biology Conference, Los Angeles, CA, 2018
8. I. Marquetti, S. Desai, Molecular modeling of bone morphogenetic protein for tissue engineering applications. Proceedings of the Industrial Engineers Research Conference, Orlando, FL, 2018
9. T. Hanawa, Evaluation techniques of biomaterials in vitro. Sci. Technol. Adv. Mater. **3**, 289–295 (2002)
10. B.D. Ratner, A.S. Hoffman, F.J. Schoen, J.E. Lemons, *Biomaterials Science: An Introduction to Materials in Medicine*, 2nd edn. (Elsevier Academic Press, San Francisco, 2004)
11. R.J. Good, Contact angle, wetting, and adhesion: A critical review, in *Contact Angle, Wettability and Adhesion*, ed. by K. L. Mittal, (VSP Publishers, Netherlands, 1993)
12. C.J. Beeves, J.L. Robnison, Some observations on the influence of oxygen content on the fatigue behavior of α-titanium. J. Less-Common Met. **17**, 345–352 (1969)
13. H. Conrad, M. Doner, B. de Meester, Critical review: Deformation and fracture, in *Titanium, Science and Technology*, ed. by R. I. Jaffee, H. M. Burte, vol. 2, (TMS, Warrendale, PA, 1973)
14. M.J. Donachie Jr., *Titanium: A Technical Guide* (ASM International, Metals Park, OH, 1988), p. 11. ISBN 0871703092
15. *Titanium. Columbia Encyclopedia*, 6th edn. (Columbia University Press, 2000–2006, New York). ISBN 0787650153
16. *Titanium. Encyclopedia Britannica,* 2006
17. M.J. Donachie Jr., *Titanium: A Technical Guide* (Metals Park, OH, ASM International, 1988)., Chapter 4. ISBN 0871703092
18. G.Z. Chen, D.J. Fray, T.W. Farthing, Direct electrochemical reduction of titanium dioxide to titanium in molten calcium chloride. Nature **407**, 361–364 (2000). https://doi.org/10.1038/35030069
19. H.J. Breme, V. Biehl, J.A. Helsen, Metals and implants, in *Metals as Biomaterials*, ed. by J. A. Helsen, H. J. Breme, (Wiley, Chichester, 1998), pp. 37–72
20. J.B. Park, Metallic biomaterials, in *The Biomedical Engineering Handbook*, ed. by J. D. Bronzino, (CRC Press, Boca Raton, 1995), pp. 537–551
21. J.E. Davies, B. Lowenberg, A. Shiga, The bone–titanium interface in vitro. J. Biomed. Mater. Res. **24**, 1289–1306 (1990)
22. N.J. Hallab, J.J. Jacobs, J.L. Katz, Orthopedic applications, in *Biomaterials Science—An Introduction to Materials in Medicine*, ed. by B. D. Ratner, A. S. Hoffman, F. J. Schoen, J. E. Lemons, (Elsevier/Academic Press, San Diego, 2004), pp. 526–555
23. A.N. Cranin, J.E. Lemons, Dental implantation, in *Biomaterials Science—An Introduction to Materials in Medicine*, ed. by B. D. Ratner, A. S. Hoffman, F. J. Schoen, J. E. Lemons, (Elsevier/Academic Press, San Diego, 2004), pp. 555–572
24. DePuy Orthopaedics Inc. http://www.jointreplacement.com/xq/ASP.default/pg.content/content_id.84/mn.local/joint_id.5/joint_nm.Hip/local_id.4/qx/default.htm
25. N. Tony, Report on stainless steel – A family of medical device materials. Business Briefing: Medical Device Manufacturing & Technology (2002)
26. H. Serhan, M. Slivka, T. Albert, S. Kwak, Is galvanic corrosion between titanium alloy and stainless steel spinal implants a clinical concern? Spine J. **4**(4), 379–387 (2004)
27. Carpenter Technology Corp, http://cartech.ides.com/datasheet.aspx?&I=101&E=6
28. ConMed Linvatec Corporation, http://www.conmed.com/products-knee-fixation.php
29. ConMed Linvatec Corporation, http://www.conmed.com/products-maninst-concept.php

30. W.J. Buehler, J.V. Gilfrich, R.C. Wiley, Effect of characteristic temperatures of thermoelastic martensitic properties of alloys near composition TiNi. J. Appl. Phys. **34**, 1475–1476 (1963)
31. L.G. Machado, M.A. Savi, Medical applications of shape memory alloys. Braz. J. Med. Biol. Res. **36**, 683–691 (2003)
32. D.E. Hodgson, M.H. Wu, R.J. Biermann, Shape Memory Alloys, in *Metals Handbook*, vol. 2, (ASM International, Metals Park, Ohio, 1990), pp. 897–902
33. H. Kyogoku, S.J. Komatsu, Japan Society of Powder and. Powder Metall. **46**(10), 1103 (1999)
34. S. Saito, T. Wachi, S. Hanada, Mater. Sci. Eng. **A161**, 91 (1992)
35. V.I. Itin, V.E. Gjunter, S.A. Shabalovskaya, R.L.C. Sachdeva, Mater. Charact. **32**, 179 (1994)
36. J.C. Hey, A.P. Jardine, Mat. Res. Soc. Symp. Proc. **360**, 483 (1995)
37. Z. Xu, C.K. Waters, G. Rajaram, J. Sankar, Preparation of porous nitinol material by hot-isostatic pressing. Proceedings of advances in materials processing for challenging environments, ASME International Mechanical Engineering Congress & Exposition, November 5–11, 2005, Orlando, Florida
38. Z. Xu, *Center for Advanced Materials and Smart Structures* (North Carolina A & T State University, Greensboro, North Carolina, USA, 2005)
39. D. Stockel, Nitinol medical devices and implants. Min. Invasive Ther. Allied Technol. **9**, 81–88 (2000)
40. A.R. Pelton, D. Stöckel, T.W. Duerig, Medical uses of nitinol. Mater. Sci. Forum **327–328**, 63–70 (2000)
41. D. Mantovani, Shape memory alloys: Properties and biomedical applications. J. Miner. Met. Mater. Soc. **52**, 36–44 (2000)
42. T.M. Duerig, A. Pelton, D. Stöckel, An overview of nitinol medical applications. Mater. Sci. Eng. A **273–275**, 149–160 (1999)
43. RX Acculink Carotid Stent System, Abbot Vascular Images courtesy of Abbott Vascular. (c) 2007 Abbott Laboratories. All Rights Reserved. http://www.abbottvascular.com/
44. F.J. Gil, J.A. Planell, Shape memory alloys for medical applications. Proc. Inst. Mech. Eng. H **212**(6), 473–488 (1998)
45. J. Stice, The use of superelasticity in guidewires and arthroscopic instrumentation, in *Shape Memory in Engineering Aspects of Shape Memory Alloys*, ed. by T. W. Duering, K. N. Melton, c. D Sto¨, C. M. Wayman, (Butterworth-Heinemann, London, 1990), pp. 483–486
46. D.W. James, High damping for engineering applications. Mater. Sci. Eng. **4**, 322 (1969)
47. Y. Sekiguci, in *Medical Applications in Shape Memory Alloys*, ed. by H. Funakubo, (Gordon and Breach Science Publishers, London, 1984), pp. 10–23
48. X.F. Zhang, A study of shape memory alloy for medicine, in *Shape Memory Alloy 86, Proceedings of the International Symposium on Shape Memory Alloys, China*, (Academic Publishers, New York, 1986), pp. 24–28
49. M. Simon, R. Kaplow, E. Salzman, D.A. Freiman, Vena cava filter using thermal shape memory alloy. Radiology **125**, 89–90 (1977)
50. G.F. Andreasen, A clinical trial of alignment of teeth using a 0.019 inch thermal nitinol wire with a transition temperature range between 31 °C and 45 °C. Am. J. Orthod. **78**, 528–536 (1980)
51. F. Miura, M. Mogi, Y. Ohura, M. Karibe, The superelastic japanese NiTi alloy wire for use in orthodontics. Am. J. Orthod. Dentofacial Orthop. **94**(2), 89–96 (1988)
52. H.S. Han, Y.Y. Kim, Y.C. Kim, S.Y. Cho, P.R. Cha, H.K. Seok, S.J. Yang, Bone formation within the vicinity of biodegradable magnesium alloy implant in a rat femur model. Met. Mater. Int. **18**(2), 243–247 (2012)
53. M. Staiger, A. Pietack, J. Huadmai, G. Dias, Magnesium and its alloys as orthopedic biomaterials: A review. Biomaterials **27**(9), 1728–1734 (2006)
54. R.K. Rude, Magnesium deficiency: A cause of heterogeneous disease in humans. J. Bone Miner Res. **13**, 749–758 (1998)

55. R. Siverman, H. Osborn, H.J. Runge, E.J. Gallagher, W. Chiang, J. Feldman, T. Gaeta, K. Freeman, B. Levin, N. Mancherje, S. Scharf, IV magnesium sulfate in the treatment of acute severe asthma: A multicenter randomized controlled trial. Chest **122**(2), 489–497 (2002)
56. H.G. Stuhlinger, Magnesium in cardiovascular disease. J. Clin. Basic Cardiol. **5**(1), 55–59 (2002)
57. J. Perkins, Z. Xu, C. Smith, A. Roy, P.N. Kumta, J. Waterman, D. Conklin, S. Desai, Direct writing of polymeric coatings on magnesium alloy for tracheal stent applications. Ann. Biomed. Eng. **43**(5), 1158–1165 (2015)
58. G. Song, S. Song, A possible biodegradable magnesium implant material. Adv. Eng. Mater. **9**(4), 298–302 (2007)
59. J.D. Santos, Ceramics in medicine, in *Business Briefing: Medical Device Manufacturing and Technology*, Greensboro, North Carolina, USA, (2002), pp. 1–2
60. A.S. Vlasov, T.A. Karabanova, Ceramics and medicine (review). Glas. Ceram. **50**(9—10) (1994)
61. L.M. Rodríguez-Lorenzo, M. Vallet-Regí, J.M.F. Ferreira, M.P. Ginebra, C. Aparicio, J.A. Planell, Hydroxyapatite ceramic bodies with tailored mechanical properties for different applications. J. Biomed. Mater. Res. **60**(1), 159–166 (2002)
62. Biomet® Hip Hydroxyapatite Joint Replacement Prostheses, Biomet Orthopedics, Inc. http://www.biomet.com/hcp/prodpage.cfm?s=0901&p=090F
63. R.C. Garvie, R.H. Hannink, R.T. Pascoe, Ceramic steel. Nature **258**(5537), 703–704 (1975)
64. M. Yildirim, H. Fischer, R. Marx, D. Edelhoff, Invivo fracture resistance of implant supported all-ceramic restorations. J. Prosthet. Dent. **90**(4), 325–331 (2003)
65. G. Kessler-Liechti, R. Mericske-Stern, Rehabilitation of an abraded occlusion with Procera-ZrO2 all-ceramic crowns. A case report. Schweiz. Monatsschr. Zahnmed. **116**, 156–167 (2006)
66. L. Rimondini, L. Cerroni, A. Carrassi, P. Torricelli, Bacterial colonization of zirconia ceramic surfaces: An in vitro and in vivo study. Int. J. Oral Maxillofac. Implants **17**(6), 793–798 (2002)
67. M. Degidi, L. Artese, A. Scarano, V. Perrotti, P. Gehrke, A. Piattelli, Inflammatory infiltrate, microvessel density, nitric oxide synthase expression, vascular endothelial growth factor expression, and proliferative activity in peri-implant soft tissues around titanium and zirconium oxide healing caps. J. Periodontol. **77**(1), 73–80 (2006)
68. Smith & Nephew Inc., Oxinium™ knee implant http://www.voteoxinium.com/1100_oxmaterial.html

Chapter 2
Emerging Trends in Polymers, Composites, and Nano Biomaterial Applications

Salil Desai and M. Ravi Shankar

2.1 Introduction

Polymers are long-chain molecules that are formed by connecting large numbers of repeating units (monomers) by covalent bonds. Polymers form the largest category of diverse biomaterials. Based on their source of origin, they can be categorized as synthetic (e.g., polyethylene) or natural (e.g., collagen). Synthetic polymers can be further subdivided into biodegradable and nondegradable types. In the degradable type, the polymer is broken down in vivo due to hydrolytic and enzymatic [1] degradation. The resultant nontoxic compounds include lactic and glycolic acid, respectively. One of the key issues while considering polymers for bioapplications is their biocompatibility with the host tissue and their degradation characteristics over extended periods of time. Biopolymer applications range from drug release carriers, implants, tissue regeneration scaffolds to sutures.

2.2 Synthetic Polymers

Synthetic polymers range from polytetrafluoroethylene (PTFE), silicon rubber, poly(methyl methacrylate) (PMMA), copoly(lactic-glycolic acid) (PLGA), polyethylene (PE), to polyurethanes. This section details specific applications of synthetic biopolymers.

S. Desai (✉)
Department of Industrial & Systems Engineering, North Carolina A&T State University, Greensboro, NC, USA
e-mail: sdesai@ncat.edu

M. R. Shankar
Department of Industrial Engineering, University of Pittsburgh, Pittsburgh, PA, USA
e-mail: ravishm@pitt.edu

P. J. Bártolo, B. Bidanda (eds.), *Bio-Materials and Prototyping Applications in Medicine*, https://doi.org/10.1007/978-3-030-35876-1_2

2.2.1 Polytetrafluoroethylene (PTFE)

Commonly known as Teflon®, polytetrafluoroethylene (PTFE) is a synthetic polymer with extremely low coefficient of friction. It exhibits hydrophobicity, nonreactive behavior, and has high elasticity. This makes Teflon® a prime candidate for the implanting of an artificial tendon or ligament in the musculoskeletal system. Its commercial woven form (ePTFE) is called GORE-TEX®. It is also used to make catheters and in facial reconstructive surgery. Figure 2.1 shows the GORE VIABAHN® endoprosthesis stent graft. Endoprosthesis is a flexible metallic tubular-shaped device which is lined with plastic (ePTFE). This device is released within a blocked femoral artery to improve blood flow.

Vascular grafts and sutures as shown in Fig. 2.2 are made from GORE-TEX® and are nonreactive and resist the spread of infection. These grafts can be designed in a wide range of configurations including straight, tapered, and bifurcated. GORE-TEX® suture is microporous and manufactured from expanded polytetrafluoroethylene (ePTFE). It also offers minimal biological tissue response with cellular ingrowth.

Hydrogels
Hydrogel is a colloidal gel in which water is the medium of dispersion and is formed by the cross-linking network of hydrophilic polymer chains [4]. Hydrogel can either be formed by chemical or physical bonds. Hydrogels containing more than 95% of water of the total weight (or volume) of the hydrogel are called superabsorbent. Hydrogels can maintain their shape due to the isotropic swelling. Some of the important properties of hydrogels for biomedical applications include their in situ formability, responsive swelling, biodegradability, and natural tissue-like properties. Hydrogels can be cross-linked using radiation and heat. They can also be

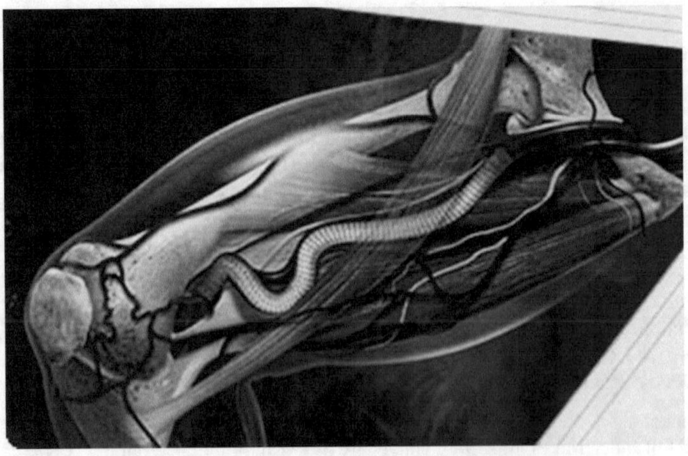

Fig. 2.1 GORE VIABAHN® endoprosthesis stent graft. (Courtesy: W. L. Gore & Associates, Inc. [2])

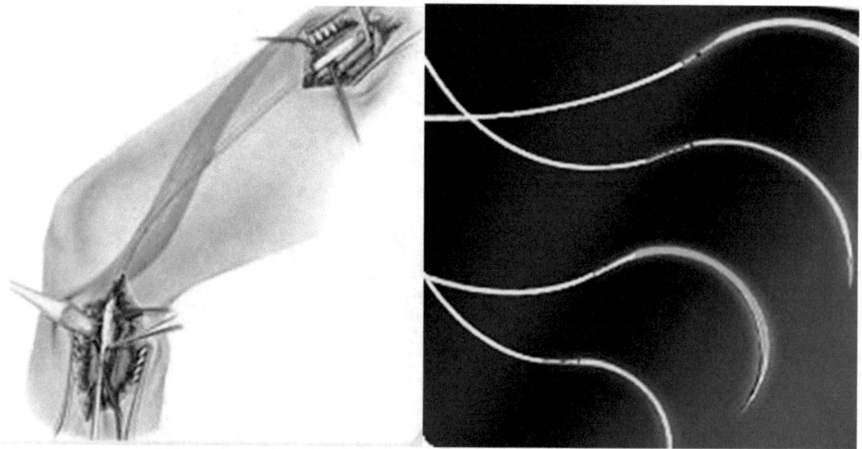

Fig. 2.2 GORE-TEX® vascular grafts (left) GORE-TEX® sutures (right). (Courtesy: W. L. Gore & Associates, Inc. [3])

degraded by the mechanism of hydrolysis and enzymatic action. They exhibit responsive swelling behavior due to changes in temperature, PH potential. The structure of hydrogels can be compared with collagen and elastin which form the natural tissue. Typical applications include controlled drug release [5], tissue regeneration scaffolds [6], cell and DNA encapsulation [7], contact lenses [8], wound healing dressings [9], and biosensors [10]. Hydrogels are formulated from a variety of materials including silicon, cellulose derivatives, poly (vinyl alcohol), poly (ethylene glycol), calcium alginate, and the most widely used poly (hydroxyethyl methacrylate) PHEMA.

Silicon hydrogel contact lenses offer superior benefits of oxygen replenishment to the cornea over traditional hydrogel soft contact lenses. Due to their high oxygen permeability, these lenses can be worn for extended periods of time over conventional hydrogel lenses. In addition, these lenses maintain hydration levels for eye comfort and have reduced occurrence of eye infections.

Figure 2.3 [11] shows microcapsules and scaffolds of calcium alginate hydrogels manufactured using specialized jetting techniques in aqueous media. Alginate is an extract of the seaweed which is used to obtain a dry, powdered sodium alginate. The sodium alginate solution is used as a precursor material to form calcium alginate microcapsules. These microcapsules have a tight control on the size distribution and can be used as drug delivery carriers. They can also be used to encapsulate cells, DNA, and biofluids. A promising application of these hydrogels is in the field of regenerative tissue engineering. Orientation-specific tissue scaffolds can be fabricated using the calcium alginate biopolymer system with different polymer loadings.

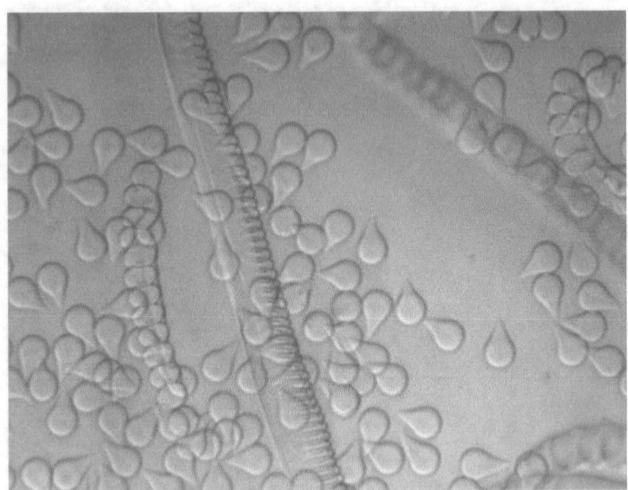

Fig. 2.3 Calcium alginate biopolymer microcapsules and tissue scaffolds in aqueous media. (Courtesy: Desai et al., Invention Disclosure, NC A&T SU [11])

2.2.2 Polymers for Dental Restorative Applications

A particularly wide-spread application of nonbiodegradable polymer systems is replacement for amalgam in dental restorations such as dental fillings. The utilization of polymer systems in dental restorations is typically determined by a) mechanical durability, b) bio-compatibility, c) ease of application, and d) aesthetic properties. The characteristics and performance of such dental restorative materials have been formalized via standards established by the International Standards Organization and the American Dental Association [12–15].

Traditionally, methacrylate-based photopolymerizable resins have been utilized since their introduction by R. F. Bowen [16, 17]. Typically, bisphenol A-glycidyl methacrylate (Bis-GMA) in combination with different fractions of triethyleneglycol dimethacrylate (TEGDMA) is photo-polymerized with visible light in the presence of suitable photo-initiators to create dental fillings. Such polymer-based fillings possess two significant advantages in comparison with traditional amalgams. First, they eliminate entirely the utilization of toxic heavy metals such as mercury in a biomedical application. Second, the optical characteristics of polymer systems enhance their aesthetic characteristics, and these restorations essentially blend in with the surrounding dental structure. By comparison, the amalgams are characterized by a typically metallic luster, and thus, they do not compare well with the polymer systems. However, currently, prevalent polymer-based dental restorations suffer significant shortcomings that have stimulated a considerable amount of research. Polymeric systems are mechanically weaker and concomitantly much less durable than, say, the traditional amalgams. In fact, it was determined that resin-based restorations typically lasted ~7.8 years in comparison with ~12.8 years for amalgam-based fillings [16].

The poor durability of the polymer systems can be traced to their intrinsically poor yield and fracture strengths, a typical characteristic of several methacrylate-based polymers. Another debilitating aspect of these polymer systems is the shrinkage that accompanies the polymerization process. Such shrinkage would inevitably lead to the formation of gaps between the restoration and the surrounding dental structure. This leads to microleakage and the deposition and entrapment of fluids, food debris, and microorganisms in the gap between the filling and the tooth. In turn, this can further engender tooth decay and overall failure of the restoration. Several developments [18] have focused on overcoming these two limitations of dental restorative polymers by developing (a) reinforced polymers for improved strength and (b) low-shrinkage polymer systems for reduced microleakage.

2.2.2.1 Reinforced Polymers for Improved Strength

The mechanical strength of polymers can be improved by reinforcing the polymer matrix via the addition of micrometer scale filler materials (e.g., zirconia, alumina, silica) [19, 20]. These micro-filled composite materials are stronger than the bulk unreinforced polymers as a result of the high-strength reinforcement phase accommodating the applied loads. This improvement in mechanical performance is much more significant when the particles are chemically treated in order to functionalize them to ensure excellent bonding between the hard reinforcement and the polymer matrix [20]. The more effective this functionalization is, the greater is the improvement in the composite strength. This is because better bonding improves the effectiveness of load transfer between the surrounding matrix and the high-strength reinforcement.

In the case of these reinforced composite, however, it was noted that the improvement in mechanical characteristics do not monotonically increase as a function of increasing fraction of the filler material [19]. In fact, it was demonstrated that there exists an optimal fraction beyond which mechanical performance such as fatigue life in fact declines in these polymer systems.

More recently, the advances in nanotechnology have led to the development of nano-filled polymers for achieving a step change in the mechanical performance of composite materials for dental restorations. The premise of these developments is that the unprecedented opportunities enabled by the novel mechanical phenomena operative at the nanometer length-scale can be exploited in suitably designed polymer matrices to create high-performance nanostructured systems. Figure 2.4 illustrates a nanostructured composite material reinforced by nano-scale silica particles.

Such nano-composites are commercially available under brand names such as Ceram X (Dentsply/Caulk, USA), Grandio (Voco, Germany), Tetric EvoCeram (Vivadent, Liechtenstein), and Filtek Supreme (3 M ESPE, USA). It should however be noted that exploiting the mechanical consequences of nanostructured dental composites still requires the rigorous development of protocols for ensuring

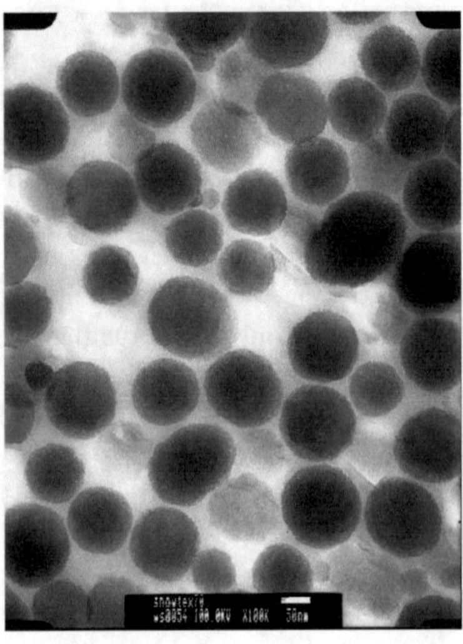

Fig. 2.4 Nanostructured dental composite reinforced by nano-scale silica particles for superior mechanical performance. (Courtesy: Elsevier [19])

effective bonding between the matrix and the nanoparticles and the generation of a fine dispersion of nanoparticles throughout the matrix. This is because of the lack of sufficient bonding between the matrix and the nanoparticle reinforcement, and concomitantly, inefficient load transfer usually results in insignificant improvement of mechanical properties over conventional materials. Instead of the high-strength, high-elastic modulus nanoparticle reinforcement accommodating the applied stresses, the poorly bonded nano-composite would behave as if it were nano-porous. This would then result in little or no improvement over the non-reinforced bulk composite material [20–22]. Furthermore, the nanoscale reinforcements are best exploited when they are uniformly dispersed in the softer matrix and there is negligible clumping or aggregation of the nano-particulates [21, 22]. Lack of aggregation is particularly important since extended ensembles of the nanoparticles would essentially behave as if they were micrometer-scale hard phases and offer little improvement over conventional polymer composites with micrometer-sized reinforcements [23].

2.2.2.2 Low-Shrinkage Polymer Systems for Reduced Microleakage

In a parallel development, to overcome the other debilitating limitation of polymeric dental restorative materials, low-shrinkage resins have been envisaged. In comparison to the methacrylate-based systems, these low-shrinkage materials are typically characterized by ring-opening polymerization reactions that involve an increase in the excluded free volume and concomitantly much smaller shrinkage. Epoxy-polyol blends that can be cured with light in the visible part of the spectrum by utilizing photo-initiators such as camphorquinone are particularly interesting for dental restorative applications. In fact, it has been suggested that such systems may offer an unprecedented combination of properties that are best suited for dental restorative applications (i.e., higher mechanical strength, rapid curing with visible light, and significantly attenuated polymerization shrinkage) [24, 25]. The polymerization shrinkage in these systems was found to be further decreased by the addition of spiro-orthocarbonates [26].

The development of nanostructured composites in combination with the design of polymer matrices that demonstrate zero or net-positive volume change during polymerization is expected to lead to the development of novel high-performance dental restorative systems by overcoming all the current limitations of prevalent methacrylate-based, micro-filled polymer materials.

2.2.3 Polymeric Biomaterials for Structural Applications

Recently, significant interest has emerged in the development of polymer-based biomaterials for the fabrication of mechanically robust implants for utilization in orthopedic surgery. Traditionally, metallic alloys have been used in orthopedic implants to utilize their high elastic modulus and materials strength in typically load-bearing applications. However, introduction of metallic systems can entail numerous effects on the overall physiology. Typically, pure metals are not sufficiently strong to be directly utilized as load-bearing biomaterials. To improve their material strength, metal alloys are utilized. For example, pure unalloyed iron does not possess sufficient material strength or the durability for utilization in biomedical implants. However, when alloyed to create stainless steel, pure iron is transformed into a much more suitable metallic system. But stainless steel contains chromium that can be gradually released into the body, thus entailing potential allergic and toxic reactions.

Furthermore, the elastic modulus of say cortical bones range from 10 to 20 GPa [27]. By comparison, the modulus of typical metallic alloys is in the range of 100–200 GPa. When materials with dramatically different properties are in intimate contact and subjected to stresses, stress concentrations that are associated with the strain mismatch can occur. Consider, for example, a Ti plate with a modulus of ~100GPa that has undergone effective osseointegration with a bone that is characterized by a modulus of ~10 GPa. When this "composite" system of the implant and the bone is deformed by a remote stress, the majority of the load would be accommodated by the stiffer Ti implant. This is particularly true if effective osseointegration has occurred, and concomitantly there is no interfacial slip between the implant and the bone.

In accordance with the Wolff's law of stress-induced bone remodeling, if a majority of the applied loads is accommodated by the implant, then over time the surrounding native bone structure may reconfigure over time and lose bone mass [28]. Furthermore, usually the implant is smaller than the surrounding bone, and at the edges, the mismatch in mechanical properties can in fact lead to strong stress concentrations. This scenario is very analogous to that observed during typical indentation scenarios when a flat punch deforms a substrate [29]. Cumulatively, these two effects can lead to secondary fractures and anomalous reconfiguration of the bone structure as a result of the highly heterogeneous stress state that is directly related to the mismatch in the mechanical properties between the implant and the native bone structure [30].

Utilization of polymer systems offers a versatile system that enables the manufacture of implants with mechanical properties that match that of the bone. Furthermore, it is known that the native bone itself is characterized by anisotropic mechanical stiffness. For example, the cortical bone in the longitudinal direction is characterized by a stiff of 17 GPa and ~13GPa in the transverse direction. Most conventional polymers however do not possess the intrinsic stiffness that matches that of the native bone. However, the primary advantage of the polymers is the control that can be exercised via controlled dispersion of second phases.

Consider the case of fiber-reinforced polymers wherein the fibers are aligned in a certain direction. Figure 2.5 illustrates carbon fiber reinforced epoxy plates that have been utilized in epiphyseal fractures [31]. In such devices, carbon fibers of exceptional intrinsic tensile stiffness can be utilized wherein mechanical properties can be tailored by controlling the orientation of the high-modulus reinforcing phase. Carbon fibers can ensure a much greater improvement in the modulus in a direction that is parallel to their predominant direction of alignment while they may not be as effective in the transverse direction. Therefore, it is then possible to tailor via appropriate processing schemes to match the anisotropic elastic properties of the bone by controlling the dispersion and the orientation of the fibers in the composite.

Much like the case of the metallic implants, polymer composite systems are also required to ensure sufficient biocompatibility. While the aforementioned carbon fiber-reinforced composites are expected to possess significant biocompatibility, in practice however, biocompatibility has been affected due to the substantial release of carbon particles into the surrounding tissues [32]. Alternatively, it has been

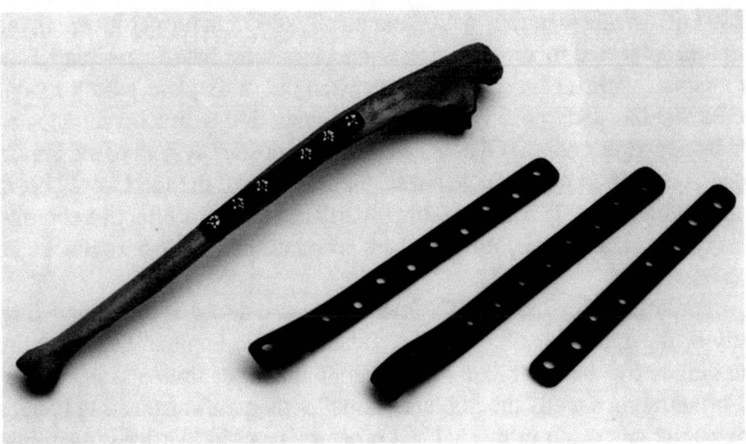

Fig. 2.5 Carbon fiber reinforced epoxy plates for epiphyseal fractures. (Courtesy: Elsevier and Orthodynamics, UK [31])

suggested that "bioglass" may be an effective reinforcement material that overcomes the limitations of the carbon fibers because they may remain bioactive even when in contact with the surrounding tissues unlike, the carbon fibers [33].

Significant challenges remain in the design and manufacture of suitable biocomposites for orthopedic and restorative applications to ensure durability and biocompatibility all the while accomplishing the desired therapeutic effect.

2.2.4 Bio-functional Ffunctional Polymers for Cardiovascular Applications

Biodegradable polymers such as polyester urethane urea (PEUU) play an important role in cardiovascular applications [34]. Synthetic polymers can be used for surface modification of metallic biomaterials to alleviate the inflammatory response of cardiovascular devices or implants placed in vivo [35–37]. Bare metal stents have been on the market for treating cardiovascular ailments including restenosis, leading to cardiac arrest or stroke-like symptoms [38, 39]. Bare metal stents are typically coated with polymers that can release therapeutic agents to minimize the proliferation of cells within the vicinity of the stent surface [40]. Several coating techniques have been implemented to deposit multilayered polymer coatings on substrates [41–43]. These include direct-write inkjet [44–47], electrospinning [45], plasma and sputter deposition [46, 47], dip and spin coating [48, 49], layer-by-layer self-assembly [50], and others. The direct-write inkjet method provides flexibility to coat complex 3D devices with biological agents embedded within polymers [51]. By manipulating the process parameters and fluid rheology, one can generate gradient patterns of polymeric coatings [52–55]. Anti-proliferative drugs such as paclitaxel (Taxol) are encapsulated within synthetic PEUU coatings [56]. The direct-write inkjet method is further employed to coat substrates used for cardiovascular stents [57]. The sustained release of the Taxol drug is intended to inhibit the growth of smooth muscle cells within the lumen of an arterial wall, thereby facilitating normal blood flow.

Figure 2.6 shows the deposition of platelets on titanium (Ti) alloy surfaces coated with different formulations of the polymer (PEUU) and Taxol drug loadings. These samples were in contact with fresh ovine (sheep) blood and exhibited variable platelet adhesion behavior. Figure 2.6a shows significant platelet deposition with pseudopodia extensions. However, samples coated with polymeric blends of Taxol drug showed marked reduction in platelet adhesion as compared with the control (PEUU coating only). The platelet counts on both Ti–PEUU–Taxol 5% and Ti–PEUU–Taxol 10% displayed a 50% reduction as compared with the Ti–PEUU control ($n = 3, p < 0.1$).

Figure 2.7 shows the relative rat smooth muscle cell (RSMC) growth inhibition on PEUU–Taxol blend multilayer coatings compared with controls. The mitochondrial activity release was tracked for days 1, 4, and 7 for the different substrate

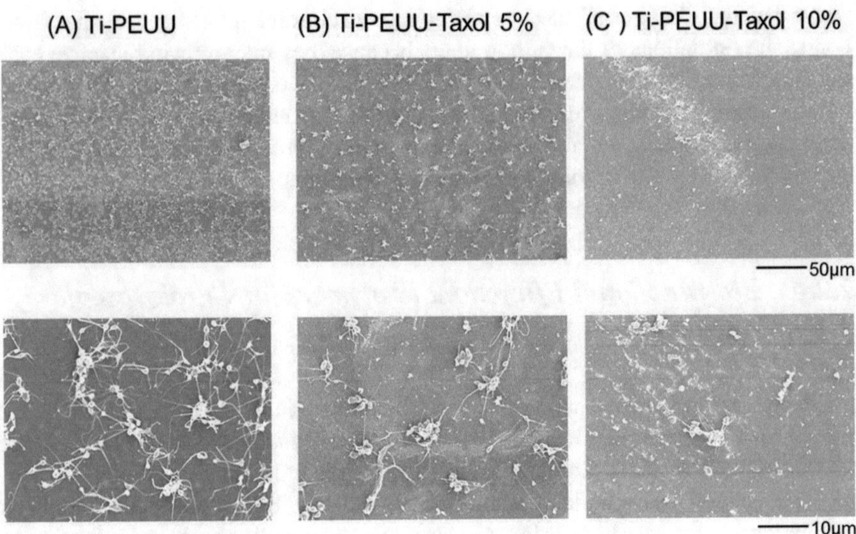

Fig. 2.6 Platelet depositions on titanium alloy surfaces coated with (**a**) PEUU no drug loading (Ti–PEUU), (**b**) PEUU coating with 5% Taxol concentration (Ti–PEUU–Taxol 5%), and (**c**) PEUU coating with 10% Taxol concentration (Ti–PEUU–Taxol 10%) [43]

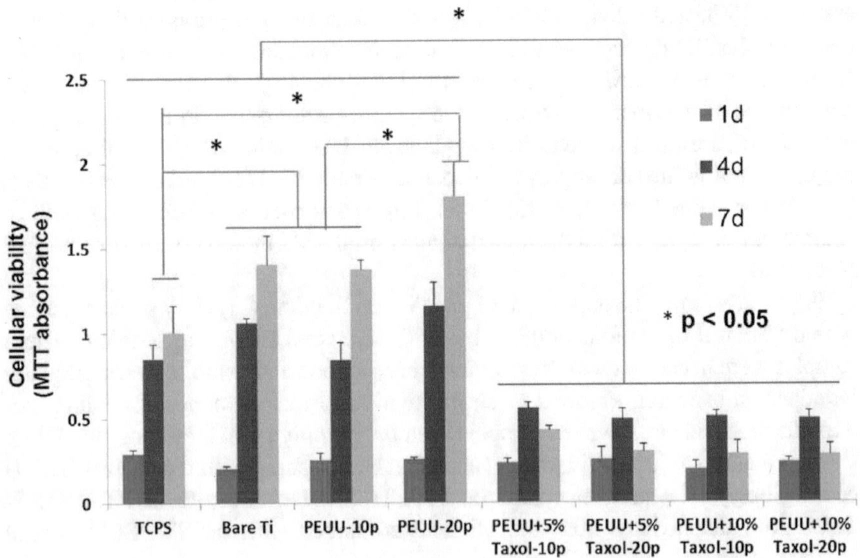

Fig. 2.7 Relative rat smooth muscle cell growth inhibition on PEUU–Taxol blend multilayer coatings compared with controls of TCPS, bare Ti, and PEUU coatings [44]

materials. The tissue culture plate substrate (TCPS) substrate represented the control surface with the baseline natural metabolic index. Bare titanium (Ti), PEUU with 10, and PEUU with 20 layers were used as positive control substrates. PEUU polymer was blended with 5% and 10% w/v Taxol drug and coated with 10 and 20 layers on Ti substrates. From Fig. 2.7, it is evident that Ti substrates coated with PEUU–Taxol drug blends had statistically lower metabolic index as compared with all the controls ($p < 0.05$). The polymer (PEUU) coatings embedded with Taxol displayed a sustained release of the drug to inhibit the RSMC proliferation. Thus, different formulations of polymers can serve as a repository for controlled release of drugs, growth factors, and other biological agents.

2.3 Nanomaterials

Materials that display structural configuration and morphology at the nanometer scale (less than 100 nm) are called nanomaterials. These include different types of materials, namely, bulk materials with nanocrystalline grain sizes, 1D nanomaterials such as thin films, 2D nanomaterials including nanowires or nanotubes, and 3D nanomaterials such as quantum dots and particulates with overall dimensions in the nanometer ranges. Nanomaterials possess unique properties over conventional materials that can be exploited for various medical applications [58–63]. Nanomaterials have potential biomedical applications that include biomarking, cancer treatment, gene therapy, biosensors, orthopedic and device implants, and targeted drug delivery [64–66]. Several nano-/microscale manufacturing processes have been implemented to fabricate biomedical coatings, devices, and implants [67–69]. These include enhanced reactivity, strength, magnetic, electrical, and optical characteristics due to size and surface effects [70–74]. Based on the material type, it is possible to alter the melting point, transparency, color, and solubility of nanoparticles by varying their particle size.

Magnetic nanoparticles are being investigated for targeting chemotherapy drugs using an external magnetic field.

2.3.1 Carbon Nanotubes (CNTs)

2.3.1.1 Description

Carbon nanotubes are cylindrical tubes of carbon with very high aspect ratios. The lengths of the carbon nanotubes are extremely large as compared with its diameter. These tubes are allotropes of carbon called fullerenes. Carbon nanotubes can either have closed or open-end configurations. Depending on their structural configuration which includes the diameter, length, and spiral twist, they possess a wide range of structural, electronic, and thermal characteristics. Carbon nanotubes are produced in two forms, namely, single-wall and multiwall nanotubes. Some of prominent characteristics of CNTs include high strength, good electrical and thermal conductivity, and functional properties [75–78].

2.3.1.2 Applications

Carbon nanotubes are being currently researched for a variety of medical applications. These include imaging, drug delivery, and biosensing for detecting disease-related proteins and chemical toxins. Carbon nanotubes can be used as structural reinforcement for tissue scaffolds. Their enhanced electrical conductivity can be exploited for directional cell growth. The different applications of carbon nanotubes in tissue engineering include cellular sensing, cell tracking and labeling, and improving tissue matrices [79]. One of the promising approaches to disease detection is the study of novel electrochemical impedance properties of carbon nanotube (CNT) tower electrode. In this method, CNT arrays are synthesized on a multilayered silicon ($Fe–Al_2O_3–SiO_2–Si$) substrate using thermal chemical vapor deposition. By controlling the process parameters, high-density and pure CNT arrays are grown which are suitable for biosensor development. Further, these CNT towers are peeled off from the silicon substrate and casted in an epoxy composite. After casting, the bottom section of the CNT tower is polished and connected to a copper wire for conductive leads. The top section of the CNT arrays is polished to expose the nanotube electrodes. These open electrodes are further functionalized by carboxylic acid groups so that they can act as receptors for other molecules. Anti-mouse IgG is covalently immobilized on the nanotube array. Finally, electrochemical impedance spectroscopy (EIS) is used to characterize the binding of mouse IgG to its specific antibody that has been immobilized on the nanotube electrode surface. Figure 2.8 shows the electrochemical impedance spectra for different activation conditions.

In recent years, engineered nanomaterials with specific chemical and physical properties are being manufactured. Given their potential use in biomedical

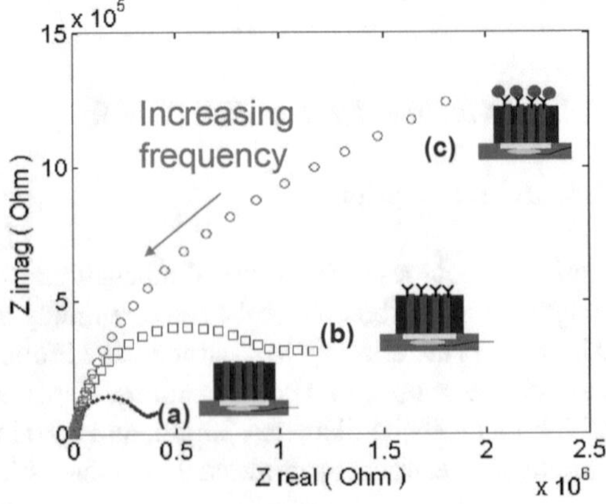

Fig. 2.8 Electrochemical impedance spectra for carbon nanotube array electrode: (a) activated nanotube array, (b) immobilized with donkey anti-mouse IgG, and (c) bound to mouse IgG at 0.2 V over a frequency range between 0.1 Hz and 300 KHz

applications, it is important to understand their toxicological effects on the human body [80]. The exposure of engineered nanomaterials during the entire lifecycle of products take a variety of routes into the body. Some of the studied exposure routes include inhalation, dermal, and oral [81, 82]. Studies have indicated oxidative stress-related inflammatory reactions due to deposition of nanoparticles within the respiratory tract after inhalation [83]. Also, the uptake of nanoparticles within the brain and gastrointestinal tract has been reported [84, 85]. However, only a limited number of materials have been studied, and further research on combinatorial nanomaterial systems needs to be explored.

Acknowledgments This work was funded by the US National Science Foundation (NSF CMMI: Award 1663128).

References

1. Y. Yaacobi, S. Sideman, N. Lotan, A mechanistic model for the enzymic degradation of synthetic biopolymers. Life Support Syst. **3**(4), 313–326 (1985)
2. W. L. Gore & Associates Inc., http://www.goremedical.com/viabahnsfa/index
3. W. L. Gore & Associates Inc., http://www.goremedical.com/vg/index. http://www.goremedical.com/suture/index
4. O. Wichterle, D. Lim, Hydrophilic gels for biological use. Nature **185**, 117–118 (1960)
5. H. He, X. Cao, L.J. Lee, Design of a novel hydrogel-based intelligent system for controlled drug release. J. Control. Release **95**(3), 391–402 (2004)
6. K.T. Nguyen, J.L. West, Photopolymerizable hydrogels for tissue engineering applications. Biomaterials **23**, 4307–4314 (2002)
7. Y. An, J.A. Hubbell, Intraarterial protein delivery via intimally-adherent bilayer hydrogels. J. Control. Release **64**, 205–215 (2000)
8. Larsen, United States Patent – 4,495,313, Preparation of hydrogel for soft contact lens with water displaceable boric acid ester, Jan 1985
9. J.A. Hubbell, Hydrogel systems for barriers and local drug delivery in the control of wound healing. J. Control. Release **39**, 305–313 (1996)
10. P. J. Gardner, A. W. Fountain III (eds.), *Chemical and Biological Sensing VII. Proceedings of the SPIE*, SPIE Press, Bellingham, Washington **6218** (2006), p. 62180K
11. Desai S, Moore A, Sankar, Invention disclosure: Method for producing uniform sized biopolymer microbeads using specialized inkjet printing, NCA&T SU: EN0046 0307, Nov 2006
12. International Standards Organization, Dental materials—water-based cements Part 1—powder/liquid acid–base cement. 2001; ISO 9917-1
13. International Standards Organization, Dentistry—resin-based filling, restorative and luting materials. 2000; ISO 4049
14. International Standards Organization, Dental water-based cements Part 2—light activated cements. 1998; ISO 9917-2
15. http://www.ada.org/prof/resources/positions/standards/denmat.asp
16. J.P. Van Nieuwenhuysen, W. D'Hoore, J. Carvalho, V. Qvist, J. Dent. **31**(6), 395–405 (2003)
17. A. Htang, M. Ohsawa, H. Matsumoto, Dent. Mater. **11**(1), 7 (1995)
18. U.J. Yap, X. Wang, X. Wu, S.M. Chung, Biomaterials **25**, 2179 (2004)
19. M. Chen, C. Chen, S. Hsu, S. Sun, W. Su, Dent. Mater. **22**(2), 138–145 (2006)
20. P.M. Ajayan, L.S. Schadler, Adv. Mater. **12**(10), 750 (2000)
21. O. Breuer, U. Sundararaj, Polym. Compos. **25**(6), 630 (2004)

22. L.S. Schadler, S.C. Giannaris, P.M. Ajayan, Appl. Phys. Lett. **73**(26), 3842 (1998)
23. Y. Ou, F. Yang, Yu ZZ. J. Polym. Sci. B Polym. Phys. **36**(5), 789 (1998)
24. A.B. Oraleg, L. Wictorin, A. Larsson, Photopolymerizable composition, Patent WO 95/30402 (1995)
25. D. Kaisaki, S. Mitra, W.J. Schultz, R.J. Devoe. Visible light curable epoxy system with enhanced depth of cure. Patent WO 96/13528, 1-49 (1996)
26. C.C. Chappelow, C.S. Pinzino, L. Jeang, C.D. Harris, A.J. Holder, J.D. Eick, J. Appl. Polym. Sci. **76**, 1715 (2000)
27. J. Black, G.W. Hastings, *Handbook of Biomaterial Properties* (Chapman and Hall, London, UK, 1998)
28. W.C. Hayes, B. Snyder, Mechanical properties of bone, The joint ASME-ASCE applied mechanics, fluid engineering and bioengineering conference, AMD, vol. 45, Boulder, Colorado, 1981
29. K.L. Johnson, *Contact Mechanics* (Cambridge University Press, Cambridge, 1985)
30. R. Huiskes, Acta Orthop. Scand. **64**(6), 699 (1993)
31. S.L. Evans, P.J. Gregson, Biomaterials **19**, 1329 (1998)
32. L. Claes, C. Burri, R. Neugebauer, U. Gruber, Experimental investigations of hip prostheses with carbon fiber reinforced carbon shafts and ceramic heads, in *Ceramics in Surgery*, (Elsevier, Amsterdam, 1983)
33. M. Marcolongo, P. Ducheyne, E. Schepers, J. Garino, The halo effect: surface reactions of a bioactive glass fiber/polymeric composite in vitro and in vivo, 5th Bio Congress, Toronto, 1996
34. J. Guan, M.S. Sacks, E.J. Beckman, W.R. Wagner, Synthesis, characterization, and cytocompatibility of elastomeric, biodegradable poly(ester-urethane)ureas based on poly(caprolactone) and putrescine. J. Biomed. Mater. Res. **61**, 493–503 (2002)
35. T. Hanawa, In vivo metallic biomaterials and surface modification. Mater. Sci. Eng. A **267**, 260–266 (1999)
36. L.R. Madden, D.J. Mortisen, E.M. Sussman, S.K. Dupras, J.A. Fugate, J.L. Cuy, K.D. Hauch, M.A. Laflamme, C.E. Murry, B.D. Ratner, Proangiogenic scaffolds as functional templates for cardiac tissue engineering. Proc. Natl. Acad. Sci. U. S. A. **107**, 15211–15216 (2010)
37. R. Kornowski, M. Hong, F. Tio, O. Bramwell, H. Wu, M. Leon, In-stent restenosis: Contributions of inflammatory responses and arterial injury to neointimal hyperplasia. J. Am. Coll. Cardiol. **31**, 224–230 (1988)
38. A. Farb, A. Burke, F. Kolodgie, R. Virmani, Pathological mechanisms of fatal late coronary stent thrombosis in humans. Circulation **108**, 1701–1706 (2003)
39. H. Schuhlen, A. Kastrati, J. Mehilli, J. Hausleiter, J. Pache, J. Dirsschinger, A. Schomig, Restenosis detected by routine angiographic follow-up and late mortality after coronary stent placement. Am. Heart J. **147**, 317–322 (2004)
40. H. Burt, W. Hunter, Drug-eluting stents: A multidisciplinary success story. Adv. Drug Deliv. Rev. **58**, 350–357 (2006)
41. R. Fattori, T. Piva, Drug-eluting stents in vascular intervention. Lancet **361**, 247–249 (2003)
42. J. Schierholz, H. Steinhauser, A.F.E. Rump, R. Berkels, G. Pulvere, Controlled release of antibiotics from biomedical polyurethanes: Morphological and structural features. Biomaterials **18**, 839–844 (1997)
43. J. Perkins, Z. Xu, A. Roy, P. Kumta, J.D. Waterman, S. Desai, Polymeric coatings for biodegradable implants. Adv. Eng. Solut. (2014)
44. J. Perkins, Y. Hong, S.H. Ye, W.R. Wagner, S. Desai, Direct writing of bio-functional coatings for cardiovascular applications. J. Biomed. Mater. Res. A **102**(12), 4290–4300 (2014)
45. X. Wang, Piezoelectric inkjet technology—from graphic printing to material deposition, in *Nanotech Conference and Expo*, (Houston, Texas, CRC Press, 2009)
46. S. Desai, A. Richardson, S.J. Lee, Bioprinting of FITC Conjugated Bovine Serum Albumin towards stem cell differentiation. Proceedings of the industrial engineers research conference, Cancun, Mexico, 2010

47. S. Desai, H. Benjamin, Direct-writing of biomedia for drug delivery and tissue regeneration, in *Printed Biomaterials*, (Springer, New York, NY, 2010), pp. 71–89
48. P. Lu, B. Ding, Applications of electrospun fibers. Recent Pat. Nanotechnol. **2**, 169–182 (2008)
49. P.K. Chu, J.Y. Chen, L.P. Wang, N. Huang, Plasma-surface modification of biomaterials. Mater. Sci. Eng. R. Rep. **36**, 143–206 (2002)
50. F. Poncin-Epaillard, G. Lageay, Surface engineering of biomaterials with plasma techniques. J. Biomater. Sci. Polym. Ed. **14**, 1005–1028 (2003)
51. H. Fang, Dip coating assisted polylactic acid deposition on steel surface: Film thickness affected by drag force and gravity. Mater. Lett. **62**, 3739–3741 (2008)
52. S. Desai, J. Perkins, B. Harrison, J. Sankar, Understanding release kinetics of biopolymer drug delivery microcapsules for biomedical applications. Mater. Sci. Eng. B **168**(1–3), 127–131 (2009)
53. S. Desai, J. Sankar, A. Moore, B. Harrison, Biomanufacturing of microcapsules for drug delivery and tissue engineering applications. Industrial engineers research conference, Vancouver, CA, 2008
54. S. Desai, A. Moore, J. Sankar, Understanding microdroplet formations for biomedical applications, in *ASME International Mechanical Engineering Congress & Exposition*, (Boston, MA, 2008)
55. J. Perkins, S. Desai, B. Harrison, J. Sankar, Understanding release kinetics of calcium alginate microcapsules using drop on demand inkjet printing, in *ASME International Mechanical Engineering Congress & Exposition*, (FL, 2009)
56. K. Norman, A. Siahkali, B. Larsen, 6 Studies of spin-coated polymer films. Annu. Rep. Sect. C **101**, 174–201 (2005)
57. J. Perkins, S. Desai, W. Wagner, Y. Hong, Biomanufacturing: Direct-writing of controlled release coatings for cardiovascular (stents) applications. Proceedings of the industrial engineers research conference, Reno, NV, 2011
58. E. Adarkwa, S. Desai, J.M. Ohodnicki, A. Roy, B. Lee, P.N. Kumta, Amorphous calcium phosphate blended polymer coatings for biomedical implants. Proceedings of the industrial engineers research conference, Montreal, Canada, 2014
59. J. Cordeiro, S. Desai, The Leidenfrost effect at the nanoscale. ASME J. Micro Nano-Manuf. **4**(4), 041001 (2016)
60. J. Cordeiro, S. Desai, The nanoscale Leidenfrost effect. Nanoscale **11**, 12139–12151 (2019)
61. T. Akter, S. Desai, Developing a predictive model for nanoimprint lithography using artificial neural networks. Mater. Des. **160**(15), 836–848 (2018)
62. A. Gaikwad, S. Desai, Understanding material deformation in nanoimprint of gold using molecular dynamics simulations. Am. J. Eng. Appl. Sci. **11**(2), 837–844 (2018)
63. J. Rodrigues, S. Desai, The effect of water droplet size, temperature and impingement velocity on gold wettability at the nanoscale. ASME J. Micro Nano-Manuf. **5**(3), 031008 (2017)
64. W. Li, B. Ruff, J. Yin, R. Venkatasubramanian, D. Mast, A. Sowani, A. Krishnaswamy, et al., Tiny medicine, in *Nanotube Superfiber Materials*, (William Andrew Publishing, Norwich, NY, 2014), pp. 713–747
65. S. Desai, M. Ravi Shankar, Polymers, composites and Nano biomaterials: Current and future developments, in *Bio-Materials and Prototyping Applications in Medicine*, (Springer, Boston, MA, 2008), pp. 15–26
66. S. Desai, B. Bidanda, P. Bartolo, Metallic and ceramic bio-materials: current and future developments, in *Bio-Materials and Prototyping Applications in Medicine*, (Springer, Boston, MA, 2008), pp. 1–14
67. S. Desai, B. Harrison, Direct-writing of biomedia for drug delivery and tissue regeneration, in *Printed Biomaterials*, (Springer, New York, NY, 2010), pp. 71–89
68. S. Desai, North Carolina A&T State University, Methods and apparatus of manufacturing micro and nano-scale features. U.S. Patent 8,573,757, 2013
69. E. Adarkwa, S. Desai, J.M. Ohodnicki, A. Roy, B. Lee, P.N. Kumta, Amorphous calcium phosphate blended polymer coatings for biomedical implants, in *IIE Annual Conference Proceedings*, Curran Press, Red Hook, NY, (2014), p. 132

70. X.Y. Qin, J.G. Kim, J.S. Lee, Synthesis and magnetic properties of nanostructured g-Ni–Fe alloys. Nanostruct. Mater. **11**(2), 259–270 (1999)
71. M. Ferrari, Cancer nanotechnology: Opportunities and challenges. Nat. Rev. Cancer **5**(3), 161–171 (2005)
72. J.K. Vasir, M.K. Reddy, V.D. Labhasetwar, Nanosystems in drug targeting: Opportunities and challenges. Curr. Nanosci. **1**(1), 47–64 (2005)
73. T.J. Webster, R.W. Siegel, R. Bizios, Osteoblast adhesion on nanophase ceramics. Biomaterials **20**(13), 1221–1227 (1999)
74. S. Iijima, Helical microtubules of graphitic carbon. Nature **354**, 56–58 (1991)
75. J.P. Lu, Elastic properties of carbon nanotubes and nanoropes. Phys. Rev. Lett. **79**, 1297–1300 (1997)
76. M.M.J. Treacy, T.W. Ebbesen, J.M. Gibson, Exceptionally high Young's modulus observed for individual carbon nanotubes. Nature **381**, 678–680 (1996)
77. M. Kociak, A.Y. Kasumov, S. Gueron, B. Reulet, I.I. Khodos, Y.B. Gorbatov, et al., Superconductivity in ropes o single-walled carbon nanotubes. Phys. Rev. Lett. **86**, 2416–2419 (2001)
78. S.N. Song, X.K. Wang, R.P.H. Chang, J.B. Ketterson, Electronic properties of grapite nanotubes from galvanomagnetic effects. Phys. Rev. Lett. **72**, 697–700 (1994)
79. B.S. Harrison, A. Atala, Carbon nanotube applications in tissue engineering. Biomaterials **28**, 344–353 (2007)
80. P.J.A. Borm, D. Robbins, S. Haubold, T. Kuhlbusch, H. Fissan, K. Donaldson, R. Schins, V. Stone, W. Kreyling, J. Lademann, J. Krutmann, D. Warheit, E. Oberdorster, The potential risks of nanomaterials: A review carried out for ECETOC. Part. Fibre Toxicol. **11**, 3 (2006)
81. G. Oberdörster, E. Oberdörster, J. Oberdörster, Nanotoxicology: An emerging discipline evolving from studies of ultrafine particles. Environ. Health Perspect. **113**(7), 823–839 (2005)
82. K. Donaldson, V. Stone, A. Clouter, L. Renwick, W. MacNee, Ultrafine particles. Occup. Environ. Med. **58**, 211–216, 119 (2001)
83. G. Oberdorster, J.N. Finkelstein, C. Johnston, R. Gelein, C. Cox, R. Baggs, et al., Acute pulmonary effects of ultrafine particles in rats and mice. Res. Rep. Health Eff. lnst. **96**, 5–74 (2000)
84. P.U. Jani, D.E. McCarthy, A.T. Florence, Titanium dioxide (rutile) particles uptake from the rat GI tract and translocation to systemic organs after oral administration. J. Pharm. **105**, 157–168 (1994)
85. J. Bockmann, H. Lahl, T. Eckert, B. Unterhalt, Titanium blood levels of dialysis patients compared to healthy volunteers. Pharmazie **55**, 468 (2000)

Chapter 3
Silk-Based Materials and Composites: Fabrication and Biomedical Applications

Golnaz Najaf Tomaraei, Se Youn Cho, Moataz Abdulhafez, and Mostafa Bedewy

3.1 Introduction

Silk is a fibrous protein-based material that is produced by over 40,000 species, including arthropods and insects such as silkworms, spiders, scorpions, mites, and bees. Among the diverse variety of silks, domesticated *Bombyx mori* (*B. mori*) silkworm silk is the most abundant natural silk and historically was used for high-quality clothing due to its superior textile properties such as lightness, fineness, pleasant feel, and unique luster [1]. Over the past several decades, silk has been of great interest to the scientific community from two main perspectives: (1) its exceptional mechanical properties and (2) its excellent biological characteristics. Mechanically, the toughness of silk is comparable with any of the best synthetic high-performance materials, including Kevlar, nylon, and high-tensile steel, as summarized in Table 3.1. From a biological standpoint, silk shows excellent biodegradability, biocompatibility, and nontoxicity [3]. Therefore, considerable effort has

G. Najaf Tomaraei · M. Abdulhafez
Department of Industrial Engineering, University of Pittsburgh, Pittsburgh, PA, USA
e-mail: gon2@pitt.edu; mma89@pitt.edu

S. Y. Cho
Carbon Composite Materials Research Center, Korea Institute of Science and Technology, Jeonbuk, South Korea
e-mail: seyoucho@kist.re.kr

M. Bedewy (✉)
Department of Industrial Engineering, University of Pittsburgh, Pittsburgh, PA, USA

Department of Chemical and Petroleum Engineering, University of Pittsburgh, Pittsburgh, PA, USA

Department of Mechanical Engineering and Materials Science, University of Pittsburgh, Pittsburgh, PA, USA
e-mail: mbedewy@pitt.edu

P. J. Bártolo, B. Bidanda (eds.), *Bio-Materials and Prototyping Applications in Medicine*, https://doi.org/10.1007/978-3-030-35876-1_3

Table 3.1 Comparison of mechanical properties for silk fibers and man-made fibers [2]

Material	Strength (GPa)	Toughness (Mj/m^3)	Extensibility
B. mori silkworm silk fiber	0.6	70	0.18
Nylon	0.95	80	0.18
Kevlar	3.6	50	0.027
High-tensile steel	1.5	6	0.008

been directed toward insights into structure, processing of silk [4–7], and reproducing the properties of natural silk fiber by reconstitution or recombination [8–10].

In recent years, the rapid evolution of technology and the increase of concern over environmental issues have opened new opportunities for silk-based materials in various research fields. Many researchers have employed silk as a promising candidate material for bio-applications such as tissue engineering and drug delivery because of the facile processes to prepare various silk-based material formats, including hydrogels, films, scaffolds, and fibers, while maintaining their excellent biological properties such as biodegradability, biocompatibility, and nontoxicity [3]. Additionally, silk-derived materials such as silk-based composites have also gained considerable attention as they enable the derivation of materials with more functionalities.

This chapter begins with a general overview of the structure and properties of silk protein obtained from *B. mori* silkworm. Next, we highlight some of the methodologies used in preparation of silk-based materials with tunable secondary structure. Then, some of the biomedical applications of silk in various material formats are reviewed. In the following section, we focus on recent attempts to design silk-derived materials in various material formats for potential biomedical applications.

3.2 *Bombyx mori* Silkworm Silk: Structure, Properties, and Processing

3.2.1 *Structure and Properties*

As shown in Fig. 3.1, native *B. mori* silk cocoon fibers are about 10–25 μm in diameter and are composed of two types of protein: two fibrils of micro-fibrous protein, silk fibroin (SF), and a glue-like protein, sericin, which holds the core microfilaments together and accounts for ~25 wt% of the total silkworm cocoon [1]. Also, raw silk fibers contain some natural impurities such as fat, waxes, inorganic salts, and coloring matters. SF can be extracted by removing sericin and impurities through a process called degumming. It involves boiling SF in an alkaline solution.

SF from *B. mori* is composed of three distinct proteins, light chain, heavy chain, and P25, which are present in a 6:6:1 ratio. The heavy chain and the light chain are linked together by a single disulfide bond [12, 13]. There are no repetitive sequences in both P25 and light chain; they consist of 220 and 244 hydrophilic amino acids (M_w ~ 25 and 26 kDa) respectively, reflecting no associations with secondary structure of SF [14, 15]. The heavy chain is composed of 5362 amino acids (M_w ~390 kDa)

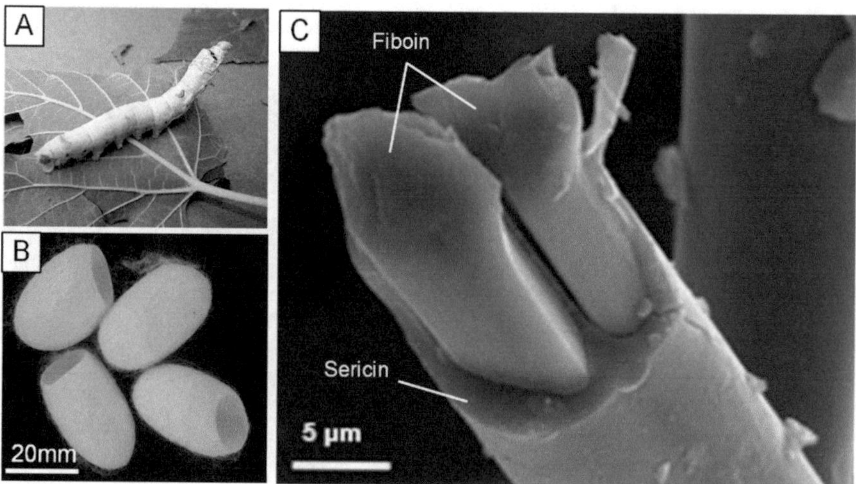

Fig. 3.1 (**a**) *B. mori* silkworm and (**b**) its cocoon fiber that consists of two fibrils of silk fibroin and a coating of sericin (**c**) as revealed by FESEM. (Reproduced from (**c**) Cho et al. [11])

mainly glycine, alanine, and serine, with a total content of above 80% [16]. The high content of these components leads to the formation of highly conserved amino acid repeat units, including GAGAGS (S is serine), GAGAGY (Y is tyrosine), GAGAGA, or GAGYGA. This promotes the construction of β-sheet crystal conformations through interchain hydrogen bonds between adjacent protein chains [17]. The amorphous regions of the heavy chain are ~40 amino acid residues in length with nonrepetitive sequence composed of charged amino acids [18, 19].

Inherently, the protein chain linked by peptide bonding is flexible to form diverse molecular conformations. However, supramolecular interactions (such as hydrogen bonding, π-interactions between aromatic groups and Van der Waals forces) result in the local conformation of polypeptide structures (secondary structure) such as α-helices, β-sheets, and turns.

Interchain hydrogen bonds between the hydrogen atom linked to the nitrogen atom and the carbonyl oxygen atom on the amino-terminal side encourage the formation of a globular structure, α-helix structure. The β-sheet conformation is organized by numerous assemblies of intra−/interchain hydrogen bonds between adjacent peptide blocks. Another secondary structure, turns, consists of four amino acids. The carbonyl oxygen of the first amino acid and the hydrogen linked to the nitrogen of the fourth amino acid form hydrogen bonding. The highly conserved primary sequence in native SF fibers determines the β-sheet-dominant secondary structure. The GX repeating units of the amino acid sequence in the crystalline domain form hydrogen bonds with each other and develop into two-dimensional chain folding, antiparallel β-sheet conformation. In addition, amino acids with small side chains such as glycine and alanine allow for inter-sheet stacking into three-dimensional nano-crystals by Van der Waals forces [2]. A typical β-sheet crystallite in SF is known to be orthogonal with the coordinate system defined with x-axis along the amino acid side chains, y-axis in the direction of the hydrogen bonds, and z-axis along the peptide bonds and with lattice constants of $a = 0.938$ nm,

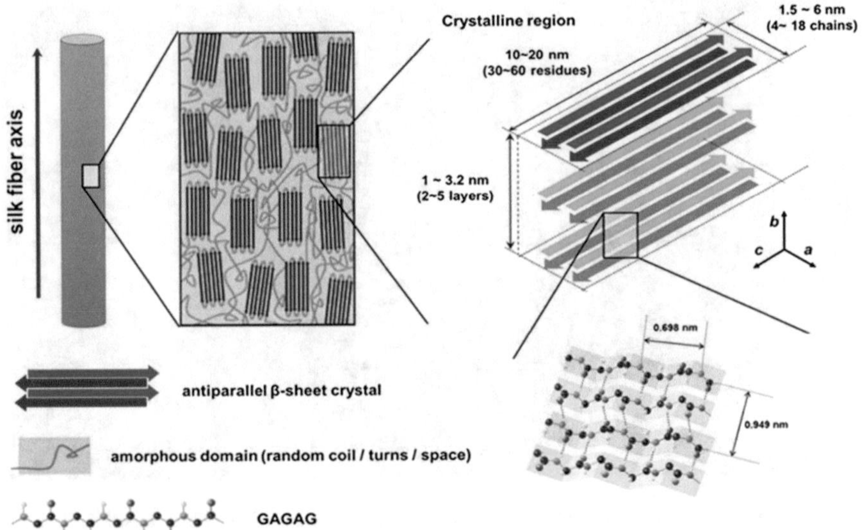

Fig. 3.2 Schematic of the hierarchical structure of β-sheet-rich silk fibroin fiber. Silk fibril shows a hetero-composite structure of stiff β-sheet crystals embedded in a soft semi-amorphous region and space. The antiparallel β-sheet crystal depicted by the ball-and-stick model (hydrogen, carbon, oxygen, and nitrogen atoms are shown as white, black, red, and blue balls, respectively). Side chains that extended above or below the sheets and alternated along the axis have been omitted for clarity

$b = 0.949$ nm, and $c = 0.698$ nm [20]. As mentioned earlier, the amorphous regions in SF are composed of non-repeated and hydrophilic amino acids. The hydrophilic amino acids are capable of hydrogen bonding with each other or with free water molecules, which enhances the flexibility of SF by acting as a plasticizer. Thus, the microstructure of native SF fibers resembles a block copolymer structure with stiff antiparallel β-sheet blocks aligned with the fiber axis and dispersed in soft amorphous segments and empty spaces as shown in Fig. 3.2. This microstructure leads to the exceptional strength while maintaining flexibility.

The β-sheet-rich structure of SF resulted from the highly conserved protein sequences results in the superior mechanical, chemical, and thermal properties of native silk fiber. Thermal and mechanical properties of SF are highly affected by absorbed water molecules below 100 °C, the boiling point of water in atmospheric pressure. In general, native silk fibers contain around 5 wt.% of water without any humidity treatment, and the absorbed water molecules are known to interact with the disordered structural components and influence the secondary structure of silk protein by acting as a plasticizer [21, 22]. The tensile strength of single fibroin fibril (prepared by degumming process) and cocoon fiber (two fibrils with sericin) shows high values in a range from 500 to 690 MPa similar to the strength of steel [23, 24]. Also, native silk fiber exhibits an excellent combination of elasticity and elongation owing to the semicrystalline structure organized by stiff β-sheets embedded in a soft semi-amorphous region. At higher temperatures above 100 °C, thermal properties of SF depend on their own chemical structure and morphologies. The onset of thermal degradation of native SF is known to be about 250 °C, which is related to the

Table 3.2 Properties of *B. mori* silkworm silk

Property		Literature
Tensile modulus (single silk fiber with sericin, GPa)	5–12	[24]
Tensile modulus (single fibril without sericin, GPa)	15–17	[23]
Tensile strength (single silk fiber with sericin, MPa)	500	[24]
Tensile strength (single fibril without sericin, MPa)	610–690	[23]
Extensibility (single silk fiber with sericin, %)	5–12	[24]
Extensibility (single fibril without sericin, %)	4–16	[23]
Glass transition temperature (°C)	178	[25]
Thermal degradation (°C)	250	[26]
Density (g/cm^3)	1.3–1.8	[27]
Crystallinity (sheet content, %)	28–62	[27]
Moisture absorption (%)	5–35	[27]

activation energy for fission of the covalent bonds in the protein's main chains [23]. The crystal melting of SF is not observed even at temperatures as high as the onset of main-chain degradation. The glass transition temperature of silk is reported around 178 °C, at which thermal and mechanical properties change dramatically [23]. In the amorphous regions, the mobility of protein chains increases prior to recrystallization, and therefore the modulus increases with temperature up to ~202 °C (Table 3.2).

Natural silkworm silk fibers possess high environmental stability due to their high crystallinity, extensive hydrogen bonds, and hydrophobic nature. As-spun silkworm silk fiber is a composite fiber of SF fibrils and sericin. On the one hand, sericin mainly consists of polar and charged amino acids and has hydrophilic properties. Thereby it's easily removed by using mild aqueous solutions under heating conditions. On the other hand, SF contains a high content of nonpolar amino acids and is insoluble in common solvents, including most alcohols and alkaline solutions, and shows only a low swelling ratio in water and dilute acids. Even strong acids such as hydrochloric acid require several hours for partial hydrolysis of SF in the amorphous regions. Also, SF shows good biodegradability and biocompatibility. Silk fiber containing highly crystallized and oriented β-sheet structure degrades slowly; however, it is possible to mediate degradation by using enzymes or controlling the content and organization of β-sheet nanocrystals [28]. As a biomaterial, SF shows good bioactive properties with similar or lower inflammatory responses compared with other biocompatible materials such as collagen and poly (lactic acid) (PLA). This was confirmed by in vitro and in vivo evaluations [29, 30].

3.2.2 *Processing of SF*

Natural silkworm silk fibers require straightforward processing for the textile industry such as dyeing and chemical modification to render colors and waterproofing properties, respectively [3]. However, to fit in a wide range of applications, it is

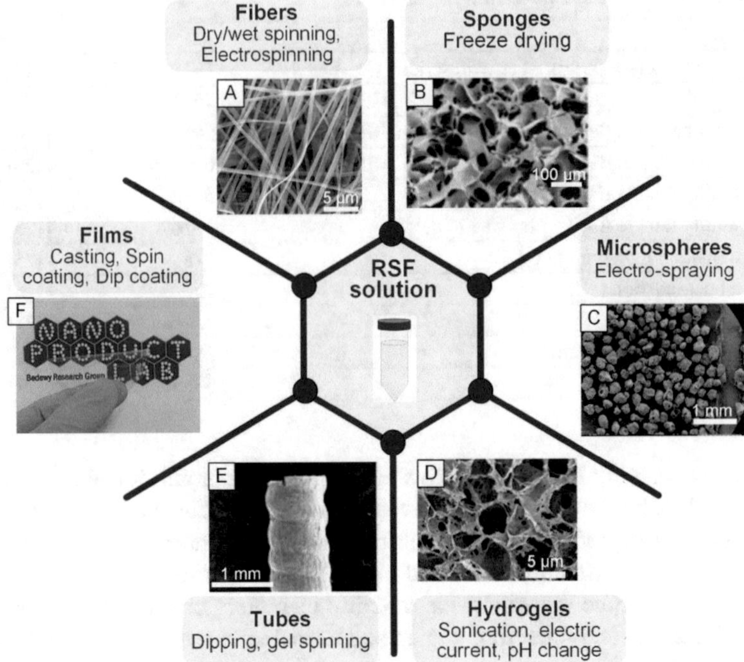

Fig. 3.3 Images of representative examples of various silk material formats and their processing methods starting from regenerated silk fibroin (RSF) solution obtained from *B. mori* silkworm silk. (**a**) Fibers, (**b**) sponges, (**c**) microspheres/microparticles, (**d**) hydrogels, (**e**) tubes, and (F) films. (Reproduced from (**a**) Sasithorn and Martinová [31]; (**b**) Tamada [32]; (**c**) Qu et al. [33]; (**d**) Kim et al. [34]; (**e**) Lovett et al. [35])

necessary to dissolve silk proteins in suitable solvents as a prerequisite step to prepare alternative material forms such as films, foams, tubes, hydrogels, fibers, and spheres as shown in Fig. 3.3. Generally, in order to break the strong intermolecular interactions between β-sheets and β-stacks in silk, concentrated aqueous solutions of inorganic/organic salts, fluorinated solvents, ionic liquids, or strong acids have been used [1, 2, 28]. Various methods have been employed to process silk into diverse material forms using regenerated SF (RSF) solution. Dry−/wet-spinning and electrospinning have been employed to create RSF fibers with micro- and nanoscale diameters, respectively. Casting, dip−/spin coating, and vacuum filtration are used to prepare RSF films. Electro-spraying and phase separation are common methods to produce RSF microspheres. Freeze-drying, gas foaming, or salt-leaching can be adjusted to prepare RSF sponges [2].

The harsh conditions during dissolving SF cause the breakage of peptide bonds in the main chain, resulting in the low crystallinity of RSF. This in turn results in inferior mechanical properties (such as brittleness and lower strength) of processed materials compared with natural SF fibers that contain high content and alignment of β-sheet crystals [36]. However, the mechanical properties of processed RSF materials can be improved by controlling the secondary structure of silk protein. This can be achieved by various treatments, which can be classified based on the processing

Table 3.3 Various treatments to tune the secondary structure of silk fibroin films

Treatment	Processing stage	Dominant resulting secondary structure	Notes	References
Formic acid dissolution	Solvent treatment	β-sheet	Requires two steps of dissolution and film casting	[38]
1,1,1,3,3,3-hexafluoroisopropyl alcohol	Solvent treatment	α-helix	Risk of chemical residues for biological application	[39]
Glycerol	Solvent treatment	α-helix	Glycerol can be totally dissolved in water	[40]
Room temp. drying	During casting	α-helix	Drying time more than 48 hrs	[41]
Casting with controlled temp. in oven	During casting	β-sheet	Slow drying in oven is needed	[42]
Water annealing	Post-processing of films	α-helix	High conc. of sol. (8 wt%) and thick film (100–200 μm) is required to have water-insoluble films	[43]
Temp. controlled water vapor treatment	Post-processing of films	Temp. dependent β content	High reproducibility Requires vacuum, Requires 12 hrs of treatment	[44]
Methanol treatment	Post-processing of films	β-sheet	Fast (10 min to 1 hour)	[38, 41]
Controlled heating	Post-processing of films	β-sheet	Transition to β-sheet starts at 140 °C	[39]

stage during which they are applied, namely, solvent treatment, treatment during processing, and post-processing treatment. For instance, post-processing exposure of silk materials to alcohols or some aqueous solutions with salts induces β-sheet formation in the materials, which enhances the mechanical properties and the stability in solvents or other environments [28, 37]. Table 3.3 summarizes some of the different treatments that have been attempted to control the secondary structure of RSF films.

3.2.3 Biomedical Applications of SF in Various Material Formats

Owing to its appealing properties such as high strength and toughness, good biocompatibility, tunable biodegradability, and permeability for oxygen and water vapor, SF in various formats is considered a potential candidate for several biomedi-

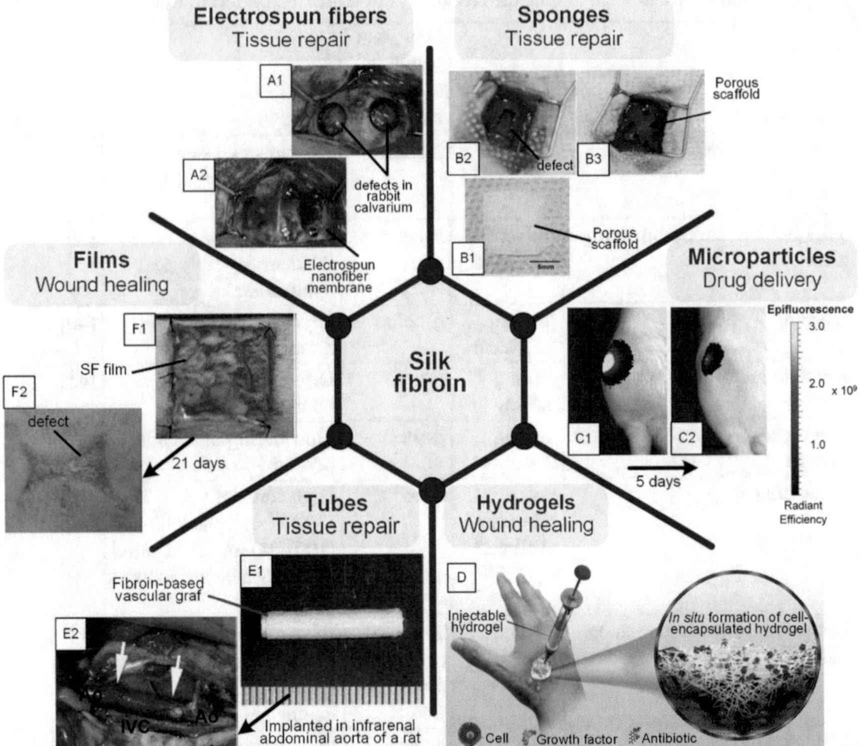

Fig. 3.4 Examples of some biomedical applications of various silk fibroin material formats. (**A1**) Defects were drilled in a rabbit calvarium model and (**A2**) the efficacy of electrospun silk membranes were evaluated in vivo. (**B1**) A porous silk fibroin scaffold to repair (**B2, 3**) a defect in the trachea of a rabbit model. (**C1**) Silk fibroin was labeled by CY7 dye as a model drug and was injected into the knee joints of rats. (C2) Persistent fluorescent after 5 days suggested slow and sustained release of the model drug. (**D**) Schematic representation of an injectable silk fibroin hydrogel that contains cells, antibiotics, and growth factor for wound healing. (**E1**) A vascular graft (**E2**) implanted inside the abdominal aorta of a rat showed blood flow without aneurysmal bulging of the graft for 1 year of study. (**F1**) Full-thickness skin defect in a rabbit model was covered with silk fibroin film, and (**F2**) defect evaluated 21 days afterwards. (Reproduced from (**A1, 2**) Kim et al. [45]; (**B1–3**) Chen et al. [46]; (**C1, 2**) Mwangi et al. [47]; (**D**) Farokhi et al. [48]; (**E1, E2**) Enomoto et al. [49]; (**F1, F2**) Zhang et al. [50])

cal applications. While in vitro cell culture studies are important to understand the behavior of SF materials in terms of cell growth and proliferation, long-term cell viability and tissue development also impact the applicability of silk-based materials in biomedical applications. Figure 3.4 illustrates representative examples of studies on in vivo behavior of various SF material forms for successful tissue reconstruction, wound healing, and small-molecule drug delivery.

3.3 Silk-Derived Materials and Composites

SF with remarkable features such as biocompatibility, tunable degradability, ease of functionalization and chemical modification, and ease of processing into various material formats has led to substantial growth in the number of studies on using SF-based materials in biomedical applications such as tissue engineering, drug delivery, and wearable implants. As mentioned earlier, while the dissolution process to prepare RSF solution makes it more versatile, it results in inferior mechanical properties. Incorporating a reinforcing agent to create SF-based composites with improved mechanical properties is a widely used approach. In addition to enhancing mechanical performance, other properties such as optical, electronic, and biological properties can also change, depending on the type of the reinforcing material. Additionally, newfound processing methods to develop novel materials have attracted considerable attention in silk research area. Among these methods are growing and assembling silk nanofibrils from RSF solution as well as SF carbonization.

The outline of this section is represented in Fig. 3.5. Here, we review studies on SF-based composites in various formats, including composite hydrogels (Fig. 3.5a), composite fibers (Fig. 3.5b), and composite films (Fig. 3.5f). We then review some of the studies on the role of SF as the reinforcing fiber (Fig. 3.5c). Hybrid nanocomposites are introduced next (Fig. 3.5d). Finally, we discuss some recent findings on carbonized silk and its application as flexible and wearable biosensors (Fig. 3.5e).

3.3.1 *Composite Hydrogels*

Hydrogels formed from a variety of natural polymers such as collagen and chitosan have been considered as potential biomaterials for tissue engineering. Particularly, intervertebral disk as well as cartilage tissue engineering have been major areas of focus due to the limited self-regeneration capacity of these tissues [53, 55]. As the favored scaffold format for tissue engineering, hydrogels have led to considerable progress in this area. The mechanical properties of natural hydrogels for tissue repair, however, must be improved so that they can withstand physiological loadings [55].

As a promising material for tissue repair, SF composite hydrogels have gained attention. Apart from enhancing the mechanical properties of SF hydrogels, other properties can also be improved in composite hydrogels. For instance, Ding et al. [51] noticed that adding hydroxyapatite to SF hydrogel improved the in vitro cytocompatibility and in vivo bone regeneration. Figure 3.6 shows that incorporating 60% hydroxyapatite nanoparticles in a SF hydrogel enhanced the rate of in vivo bone reconstruction when injected into model bone defects of rats.

As another example, incorporating conductive fillers such as metallic nanoparticles has been attempted to introduce interesting functionalities in silk protein.

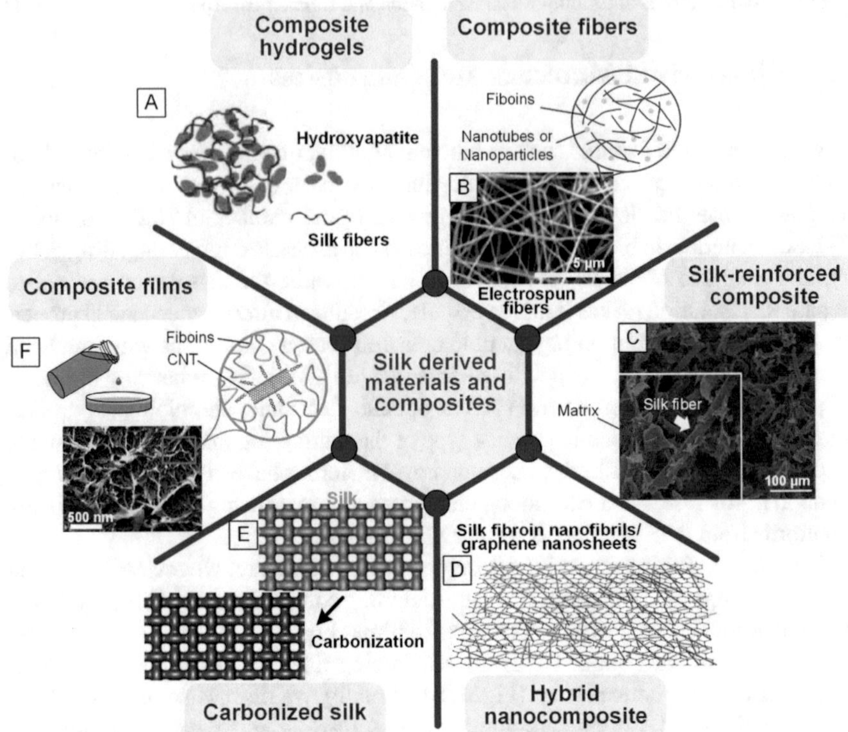

Fig. 3.5 Representative examples of silk-derived materials, including silk-based composites in various formats, that is, (**a**) hydrogels, (**b**) fibers, (**f**) films, (**c**) silk-fiber-reinforced composites, (**d**) nanohybrids containing silk fibroin nanofibrils and (**e**) carbonized silk. (Reproduced from (**a**) Ding et al. [51]; (**b**) Pan et al. [52]; (**c**) Yodmuang et al. [53]; (**d**) Ling et al. [77]; (**e**) Wang et al. [54]; and (**f**) Cho et al. [17])

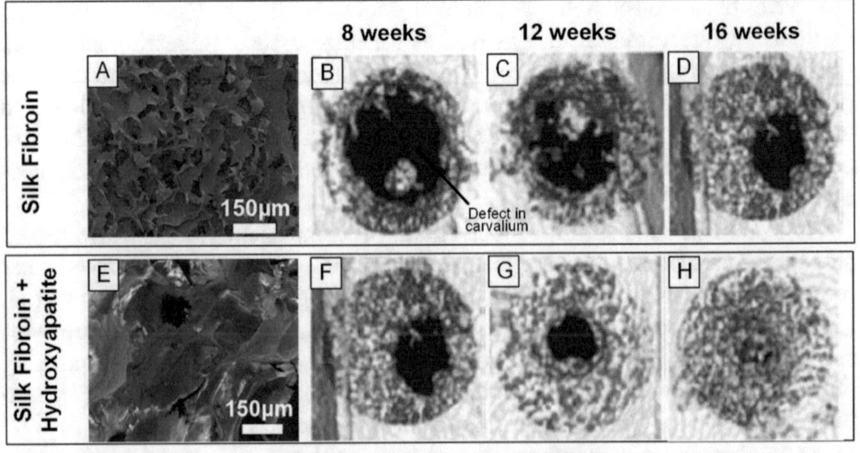

Fig. 3.6 (**a**) SEM images of the structure of (**a**) a silk fibroin hydrogel and (**e**) a silk fibroin–hydroxyapatite composite hydrogel. These hydrogels were injected into a cranial defect within a rat calvarium, and micro-CT images of the skull were used to assess the in vivo bone regeneration (**b** and **f**) 8, (**c** and **g**) 12, and (**d** and **h**) 16 weeks after injection of the hydrogels. (Reproduced from Ding et al. [51])

Gogurla et al. [56] prepared nanocomposites of SF and Au nanoparticles in the form of hydrogels by mixing silk solution and gold(III) chloride trihydrate ($HAuCl_4.3H_2O$) solution at different ratios and aging the mixtures for 15 days at room temperature. They observed negative photoconductive response for the nanocomposites, and they noticed that increasing the concentration of Au nanoparticles enhanced the electrical conductivity and photoconductance yield of the nanocomposites. These novel features of Au-silk nanocomposite hydrogels suggest possible bioelectronic and biophotonic applications such as optical biosensors.

3.3.2 *Composite Fibers*

SF in the form of electrospun nanofibers possesses high surface-area-to-volume ratio, high porosity, and a wide range of pore size and therefore is a favorable format for cell attachment and proliferation [45]. However, the physical and mechanical properties of electrospun nanofibers from aqueous solutions often need enhancement for practical use in tissue engineering applications. Blending SF with other polymers such as polyethylene oxide [57] and chitosan [58] has been observed to enhance the physical properties of nanofibers. In this regard, Gong et al. [59] pre-

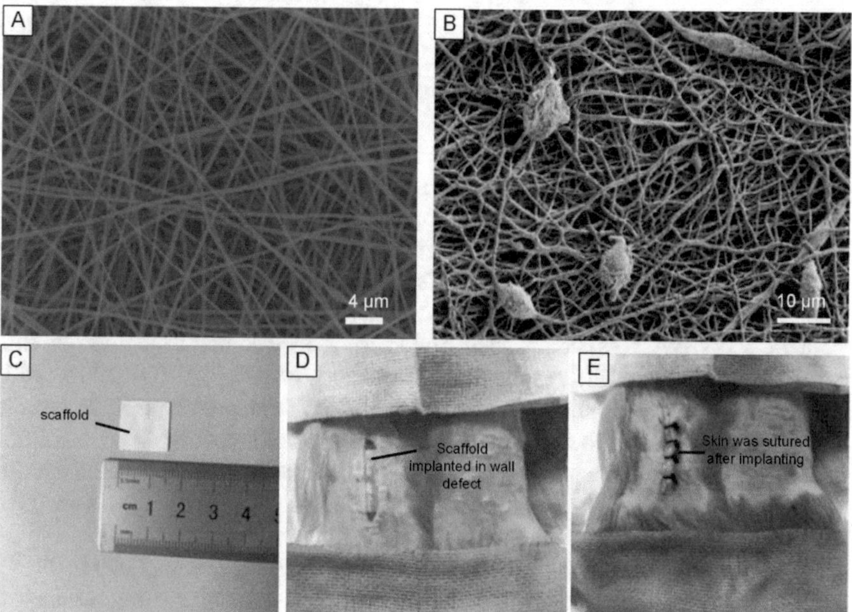

Fig. 3.7 (**a**) SEM image of electrospun composite nanofibers of silk fibroin + poly(3-hydroxybutyrate-co-3-hydroxyvalerate), PHBV, with silk fibroin to PHBV ratio of 75/25. (**b**) FESEM image of HFF-1 fibroblasts cultured on electrospun composite 75/25 nanofibers. (**c**) An electrospun scaffold was prepared by this composition, and (**d**) it was implanted to repair the abdominal wall defect in a rat. (**e**) Then, the skin incision was sutured, and the implant was kept for 15 days. (Reproduced from Gong et al. [59])

pared electrospun nanofibrous scaffolds from blends of SF and poly(3-hydroxybutyrate-co-3-hydroxyvalerate), PHBV, which is a biological polyester. As shown in Fig. 3.7, both in vitro and in vivo experiments were used to evaluate the efficacy of the scaffold in tissue regeneration. Their results showed the effectiveness of the scaffold in abdominal wall defect repair [59].

Apart from blending with other polymers to combine the desired properties of both components, a major approach to improve mechanical properties of RSF fibers is the addition of reinforcing agents such as carbon nanotubes (CNTs) into the spinning dope and making composite mats. Pan et al. [52] added functionalized multiwalled CNTs to RSF solution and fabricated electrospun composite nanofibers. Adding functionalized multiwalled CNTs up to the agglomeration point induced more β-sheet transformation of SF and improved mechanical properties. Furthermore, results indicated biocompatibility for the growth and proliferation of 3T3 cells and Begal's lingua mucosa cells.

3.3.3 Composite Films

In spite of recent advances in modern signal processing techniques, the extensive use of bioelectronic devices still requires applicable materials at the interface of the device and the biological surface [60]. The ever-increasing demand for wearable and flexible bioelectronic devices has led to considerable advances in biopolymers and carbon-based materials [61]. Natural materials such as SF have been investigated in this regard. The fact that SF is nonconductive is advantageous to its application as a dielectric component. However, it eliminates such applications where SF can be used as a direct interface between the device and biological surface, for instance, to control the bioelectrical activity of cells in neuro-regenerative medicine [61]. Therefore, with the aim of expanding the biomedical applications of SF, studies have been done on making composites of silk and conductive nanomaterials such as metallic nanoparticles and CNTs. In this regard, Dionigi et al. [61] used template method to fabricate porous nanostructured conductive films of SF + single-walled CNTs (SWCNTs) composites. Their results showed that the composite film enabled adhesion and differentiation of dorsal root ganglion (DRG) neuron cells in vitro (Fig. 3.8a). In another study, Liu et al. [64] fabricated layered composite films made of an Ag nanowire layer and a silk nanofibril layer using vacuum filtration. Good conductivity and transparency of the composite film made it suitable for designing wearable health-monitoring systems, for example, in cases where a pressure sensor with low detection limit and high response is required.

While making conductive films considerably expands the application of SF films, to meet the requirements of tissue engineering applications, the mechanical performance of RSF films must also be improved. Wang et al. observed that composite films containing SF, glycerol, and low contents of graphene oxide (up to 1%) possessed improved mechanical properties. L-929 fibroblast cells were cultured on these composite films for 7 days. SEM images showed that the cells adhered to the

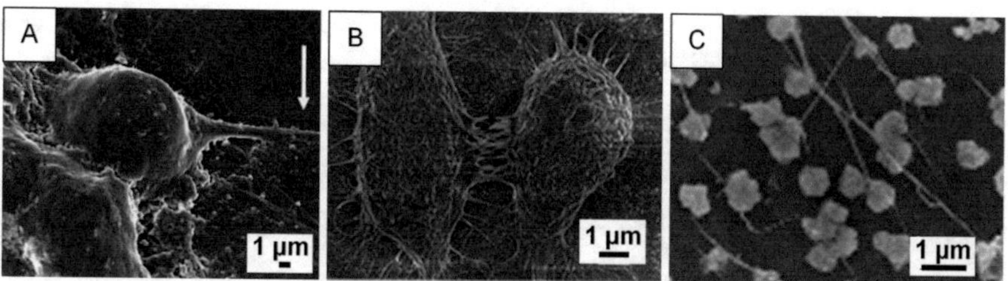

Fig. 3.8 Examples of studies on silk fibroin-based composite films by in vitro cell culturing to assess biocompatibility and cytotoxicity of the composite films. (**a**) Rat DRG neuron cells were grown on a silk fibroin-CNT composite film; SEM image showed the projection from cell body of neurons (white arrow). (**b**) L-929 cells were cultured on silk-graphene oxide composite films. SEM image after 7 days showed that cells tightly adhered to the film and were coated by extracellular matrix. (**c**) Human mesenchymal stem cells (hMSCs) were cultured on silk–silica composite films. SEM image showed mineral deposits on the surface. The mineral nodules come from bone apatite formation, while fibrous structure is due to collagen deposits. (Reproduced from (**a**) Dionigi et al. [61]; (**b**) Wang et al. [62]; (**c**) Mieszawska et al. [63])

films and were covered with extracellular matrix (Fig. 3.8b). Similarly, Mieszawska et al. [63] reported that silica induced good cell viability and proliferation of human mesenchymal stem cells (hMSCs) cultured on the composite films of SF and silica particles. Figure 3.8c shows mineral nodules connected by a collagen-based fibrillar structure after 10 weeks of cell culture on silk-2 μm silica composite films. As another reagent, adding cellulose nanofibrils (CNFs) to regenerated SF is a facile and effective method to improve mechanical properties as well as cell adhesion and proliferation [65].

It is worth noting that a different approach to adding functionality to SF films has been offered by some researchers. Hwang et al. [66] used silk film as the packaging material to seal transient implantable electronic devices which were fabricated on a silk film substrate. Along this line, Min et al. [60] showed that burying a patterned network of randomly connected silver nanowires just beneath the surface of an SF film created a flexible electrode with smooth surface, as a promising alternative for indium tin oxide (ITO) in bioelectronic devices.

In addition to nanowires, carbon nanotubes have been used as additives to SF films along with microwave processing to control the secondary structure of SF films. Cho et al. [17] created RSF–CNT composite films which were then irradiated with microwaves, resulting in local heating and local structural transition of amorphous silk fibroin molecules to stable helical structures. Upon microwave irradiation, a significant increase of helix secondary structure was observed with increasing CNT content. This approach allows a high degree of tailoring the properties and structure of functional regenerated SF-based films for flexible electronics and biomedical devices.

Table 3.4 Comparison of mechanical properties of natural and synthetic fibers used in fiber-reinforced polymer composites [27, 68]

Material	Density (g/cm^3)	Tensile strength (MPa)	Young's modulus (GPa)	Elongation at break (%)
B. mori silk	1.3–1.8	600	6–20	18
Flax	1.3	350–1050	50	3
Hemp	1.48	690	70	1.6
Jute	1.3	400–700	25	1.5
E-glass	2.7	1200	70	2.8
Carbon	1.8	4000	130	2.8

3.3.4 Silk-Fiber-Reinforced Composites

Bio-composites reinforced by natural fibers have been widely studied as potential eco-friendly alternatives to traditional fiber-reinforced composites [67]. On the one hand, traditional reinforcing fibers such as E-glass, flax, and hemp usually have a lower strain at failure and higher stiffness compared with the matrix. On the other hand, as shown in Table 3.4, *B. mori* silk fibers have a relatively low stiffness, about 6–20 GPa, and high strain at failure around 18%. Namely, silk fibers are more ductile compared with traditional polymers such as epoxy and polyester. As a result, silk-fiber-reinforced polymer composites can be described as a ductile fiber and brittle polymer matrix system, in which the toughening mechanism can be fibers acting as crack stoppers.

The technical advantage of silk fibers specific to composite applications is their naturally continuous length and the high compressibility of silk [27]. Silk is the only natural fiber existing as a filament, and thus a high fiber length distribution for effective reinforcing can be achieved by a simple process. Shah et al. [69] studied fiber volume fraction at a compaction pressure of 2.0 bar. Their results revealed that silk was significantly more compactable compared with plant-based fibers and glass fiber. They suggested that silk is suitable for fiber-reinforced composites with high fiber volume fraction.

B. mori silk fibers with a length of 0.5–10 mm were employed as reinforcements for biodegradable polymers such as poly(lactic acid) (PLA), polyurethane, and polybutylene succinate [70–74]. Manjula et al. reported that a small amount of silk fiber (less than 10 wt%) enhanced the mechanical properties and thermal stability of biopolymer. The enhancement was due to good physical interactions between ester-linkage-based biodegradable polymers and silk that consists of various functional groups, assigning polar properties. Zhao et al. prepared silk fiber/PLA bio-composites and reported that water molecules absorbed on silk fiber surface increased the free volume within PLA matrix, facilitating the ester bonds scission during enzymatic degradation measurement. They also concluded that 5 wt% was an optimum content of silk fiber in PLA matrix.

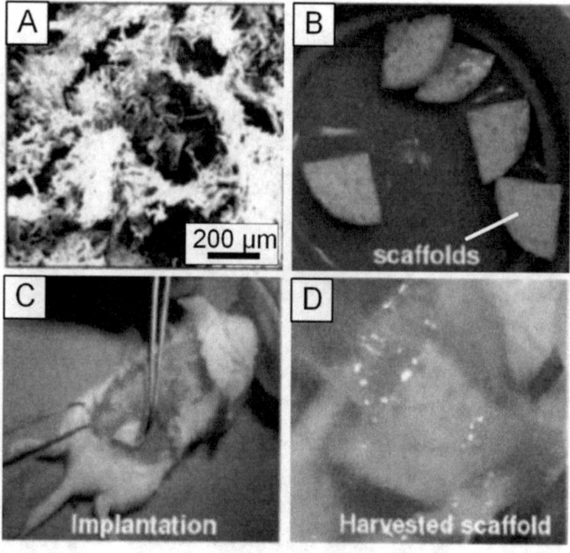

Fig. 3.9 (**a**) SEM image of the structure of a silk-microfiber-reinforced silk porous scaffold. (**b**) Optical photograph of these scaffolds. (**c**) In vivo examination of the scaffolds implanted subcutaneously at the back of mice. After 4 weeks, (**d**) dense tissue ingrowth with vascularization was observed around the implant. (Reproduced from Mandal et al. [75])

In addition to extensive studies on silk-fiber-reinforced polymers, a newer approach has been to fabricate silk-fiber-reinforced SF matrix [53]. Among natural hydrogel materials, tunable degradability of SF makes it excellent for in vivo applications [53]. Additionally, strong interfacial bonding between the two silk phases is beneficial for mechanical properties as the contribution of fibers to matrix reinforcement depends on the strength of interfacial bonding between the two. However, improving mechanical properties of SF hydrogels through fiber reinforcement requires knowledge on tissue development behavior of the composite system. Therefore, Mandal et al. [75] fabricated SF microfiber-reinforced SF hydrogels and implanted them into the back of some mice. After 4 weeks, they observed that the retrieved implants were surrounded by a dense tissue and that the implants and mice tissue were closely integrated (Fig. 3.9).

3.3.5 Hybrid Nanocomposites

As the basic architectural component of SF, silk nanofibrils (SNFs) can be grown and assembled from regenerated SF solution. Novel hybrids can be made by mixing functional nanomaterials such as silver nanowires or nanohydroxyapatite with SNFs [64, 76]. Liu et al. [64] performed vacuum filtration on a mixture of SNFs and Ag nanowire solutions to fabricate hybrid films with humidity-sensitive conductivity in various shapes. This feature along with good biocompatibility and skin affinity of

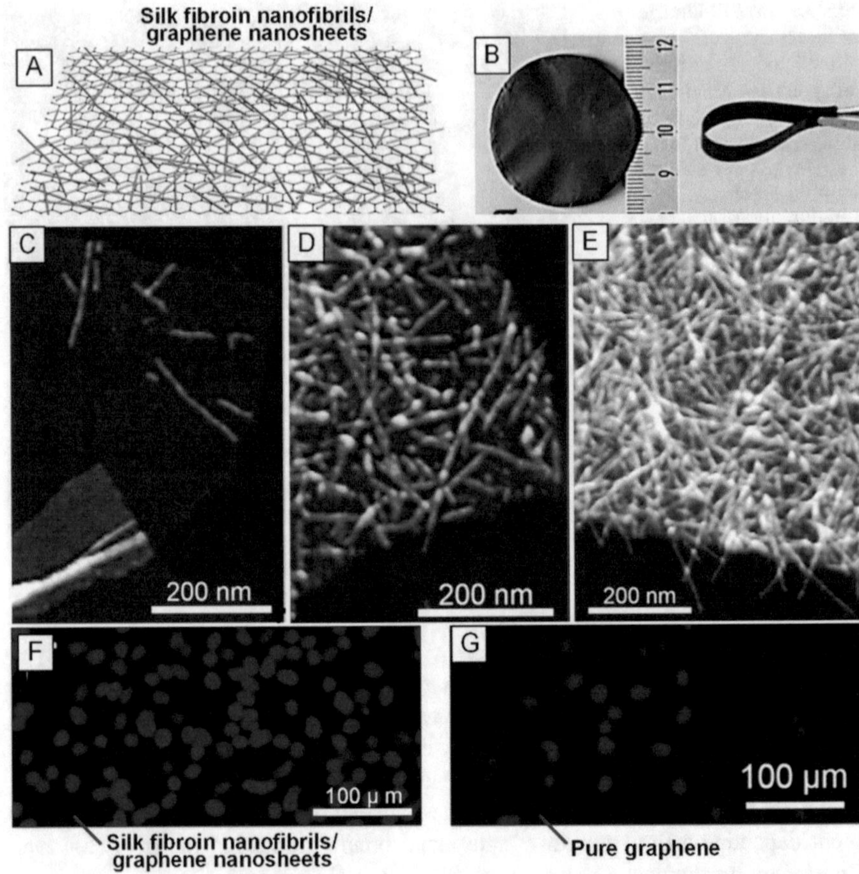

Fig. 3.10 (**a**) Schematic representation of silk nanofibrils growth on graphene nanosheets, which is formed under certain conditions of mass ratio and pH. (**b**) A composite film made by vacuum filtration of the solution mixture. The effect of mass ratio of silk nanofibers to graphene nanosheets on the hybrids was revealed by AFM at pH = 10.3. Nanofibril/graphene mass ratios were (**c**) 3/7, (**d**) 5/5, and (**e**) 8/2. The growth of stained Hela cells on the surface of hybrid film was monitored for 3 days. (**f**) The fluorescent area of the 8/2 hybrid film was about five times larger than (**g**) that for pure graphene, showing better cell adhesion and growth for hybrids. (Reproduced from Ling et al. [77])

SNFs make these hybrid films suitable for wearable health monitoring devices. Ling et al. [77] grew SNFs on graphene nanosheets (Fig. 3.10a) under different processing conditions (Fig. 3.10c–e) and fabricated nanocomposite films (Fig. 3.10b) with different ratios between SNFs and graphene nanosheets. Along with other physical and mechanical properties, they evaluated cell adhesion and proliferation of Hela cells. The film with SNF to graphene ratio of 8:2 showed better cell growth compared with a pure graphene film after 3 days of cell culture (Fig. 3.10f and g).

3.3.6 Carbonized Silk

SF as a carbon precursor has been reported recently [11, 78–80]. Generally, β-sheet-rich proteins such as silk protein can be transformed into hexagonal carbon structure through a simple thermal treatment under an inert atmosphere. Upon heating silk protein from 250 °C (the onset of thermal degradation) to 2800 °C, major chemical change and dramatic increase in conductivity occur between 300 °C and 350 °C. The carbonization of silk fibers at 800 °C was investigated, and pseudo-graphitic carbon was obtained with a carbon yield of 30% which is a higher yield than cellulose-based carbon. Further heating will result in a more developed graphitic structure. Both native silkworm silk fiber and RSF materials transform into carbon materials by heat treatment under inert atmosphere without stabilization step in air [11]. Good performance of carbonized silkworm cocoon and RSF has been reported in energy storage and conversion applications.

Flexible wearable strain sensors and skin-like pressure sensors with an electrically conductive sensing element such as metal- or carbon-nanomaterials and a flexible substrate have drawn great interest. The main challenges include finding cost-effective and environmentally friendly ways of fabricating sensors with both high sensitivity and a broad range of sensing [54, 81, 82]. Carbon materials derived

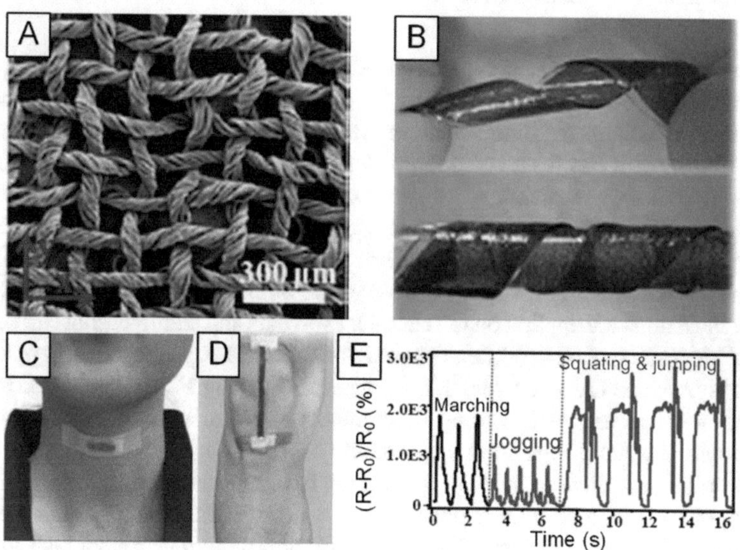

Fig. 3.11 (**a**) The SEM image of a carbonized silk fabric used to make (**b**) highly flexible strain sensors with high sensitivity and strain range. These sensors can be attached to different parts of the body, for example, (**c**) the throat or (**d**) the knee joint, to detect wide range of human activities. Different movements of the knee joint can be tracked by (**e**) recording the relative change in the resistance of the sensor over time. (Reproduced from (**a**, **b**, **d**, **e**) Wang et al. [81]; (**c**) Wang et al. [54])

from biomaterials such as silk have interesting features to be utilized in electronic devices. Unlike CNTs and graphene, carbon materials derived from biomaterials have the requirements to overcome the abovementioned challenges. Wang et al. [54, 81] made wearable strain sensors based on carbonized silk fabric and showed that the woven structure of silk fabric was well reserved after carbonization (Fig. 3.11a). Bulk shape stability had previously been reported by Liang et al. [78] during carbonization of silk cocoon. Wang et al. [54, 81] reported the influence of the woven structure of initial silk fabric on the performance of the sensor such as its sensitivity, response time, and detection limit. Figure 3.11b shows a strain sensor made by encapsulating the carbonized silk fabric within PDMS. These sensors are capable of detecting a wide range of human motions from subtle muscle movements during speech (Fig. 3.11c) to vigorous movements of knee joint (Fig. 3.11d). The sensor can differentiate various movements of the knee joint based on the relative change in the resistance of the sensor (Fig. 3.11e). Carbonized silk nanofiber membrane has also been utilized in pressure sensors. Wang et al. [82] investigated the performance of the sensors based on varying the electrospinning time during the preparation of silk nanofiber membranes, which resulted in carbonized silk nanofiber membranes with various thicknesses. The sensors made from thinner membranes showed higher sensitivities.

3.4 Summary and Outlook

Considerations regarding sustainable and environmentally friendly production have led to substantial developments in new manufacturing techniques to utilize natural materials, especially for biomedical applications. In this regard, silk fibroin from *B. mori* silkworm has been considered as a promising material due to their abundance, remarkable mechanical properties, and the facile processing techniques associated with controlling both its macroscopic form and micro-to-nanoscale morphology. At the heart of these simple processing techniques lies the RSF solution that makes silk material formats so diverse. For biomedical applications, however, more restrictions are applied on the chemical solvents and additives that can be used during processing of silk. While SF-based materials have opened several opportunities in biomedical field, considerable amount of research has also been dedicated to finding new approaches to adding functionality while maintaining its remarkable biocompatibility and tunable biodegradability. Interestingly, these approaches are quite similar regardless of the material format. In this chapter, we explain traditional mixing of SF with a secondary phase to make silk-derived composites in the form of hydrogels, fibers, and films. Then, we point at a more recent approach to grow and assemble silk nanofibrils from RSF solution and exploit them in novel nanocomposites. While the studies reviewed in this section deal with formation of silk nanofibrils, there are recent studies that try to obtain other stable building blocks such as silk nanoribbons through a proper choice of solvent system. Therefore, a potential future research direction can be aimed at the development of hybrid nanocomposites based on silk nanoribbons and other controlled morphologies. Finally, we introduce carbonized silk protein with the capability of rendering silk electrically conductive, while preserving its original form. While research on carbonized silk cocoon/fibroin continues, review of recent literature reveals that some researchers are looking at the carbonization of silk coated with nanoparticles or some polymers such as polypyrrole. In summary, research on silk fiber has been growing fast, and there is more room to be explored in this area in order to realize many biomedical technologies.

References

1. G. Chen, J. Guan, T. Xing, X. Zhou, Properties of silk fibers modified with diethylene glycol dimethacrylate. J. Appl. Polym. Sci. **102**, 424–428 (2006). https://doi.org/10.1002/app.24064
2. J.G. Hardy, L.M. Römer, T.R. Scheibel, Polymeric materials based on silk proteins. Polymer (Guildf) **49**, 4309–4327 (2008). https://doi.org/10.1016/j.polymer.2008.08.006
3. B. Kundu, R. Rajkhowa, S.C. Kundu, X. Wang, Silk fibroin biomaterials for tissue regenerations. Adv. Drug Deliv. Rev. **65**, 457–470 (2013). https://doi.org/10.1016/j.addr.2012.09.043
4. H.J. Jin, D.L. Kaplan, Mechanism of silk processing in insects and spiders. Nature **424**, 1057–1061 (2003). https://doi.org/10.1038/nature01809
5. S. Keten, Z. Xu, B. Ihle, M.J. Buehler, Nanoconfinement controls stiffness, strength and mechanical toughness of B-sheet crystals in silk. Nat. Mater. **9**, 359–367 (2010). https://doi.org/10.1038/nmat2704
6. F.G. Omenetto, New opportunities for an ancient material. Science **329**, 528–531 (2011). https://doi.org/10.1126/science.1188936
7. G.R. Plaza, P. Corsini, E. Marsano, et al., Old silks endowed with new properties. Macromolecules **42**, 8977–8982 (2009). https://doi.org/10.1021/ma9017235
8. S. Ling, Z. Qin, C. Li, et al., Polymorphic regenerated silk fibers assembled through bioinspired spinning. Nat. Commun. **8**, 1387 (2017). https://doi.org/10.1038/s41467-017-00613-5
9. F. Teule, Y.-G. Miao, B.-H. Sohn, et al., Silkworms transformed with chimeric silkworm/spider silk genes spin composite silk fibers with improved mechanical properties. Proc. Natl. Acad. Sci. **109**, 923–928 (2012). https://doi.org/10.1073/pnas.1109420109
10. G. Zhou, Z. Shao, D.P. Knight, et al., Silk fibers extruded artificially from aqueous solutions of regenerated bombyx mori silk fibroin are tougher than their natural counterparts. Adv. Mater. **21**, 366–370 (2009). https://doi.org/10.1002/adma.200800582
11. S.Y. Cho, Y.S. Yun, S. Lee, et al., Carbonization of a stable β-sheet-rich silk protein into a pseudographitic pyroprotein. Nat. Commun. **6**, 7145–7151 (2015). https://doi.org/10.1038/ncomms8145
12. T. Asakura, R. Sugino, T. Okumura, Y. Nakazawa, The role of irregular unit, GAAS, on the secondary structure of Bombyx mori silk fibroin studied with 13C CP/MAS NMR and wide-angle X-ray scattering. Protein Sci. **11**, 1873–1877 (2002). https://doi.org/10.1110/ps.0208502
13. S. Inoue, K. Tanaka, F. Arisaka, et al., Silk fibroin of Bombyx mori is secreted, assembling a high molecular mass elementary unit consisting of H-chain, L-chain, and P25, with a 6:6:1 molar ratio. J. Biol. Chem. **275**, 40517–40528 (2000). https://doi.org/10.1074/jbc.M006897200
14. M. Chevillard, P. Couble, J.-C. Prudhomme, Complete nucleotide sequence of the gene encoding the Bombyx mori silk protein P25 and predicted amino acid sequence of the protein. Nucleic Acids Res. **14**, 6341–6342 (1986)
15. K. Tanaka, S. Inoue, S. Mizuno, Hydrophobic interaction of P25, containing Asn-linked oligosaccharide chains, with the H-L complex of silk fibroin produced by Bombyx mori. Insect Biochem. Mol. Biol. **29**, 269–276 (1999). https://doi.org/10.1016/S0965-1748(98)00135-0

16. A.R. Murphy, D.L. Kaplan, Biomedical applications of chemically-modified silk fibroin. J. Mater. Chem. **19**, 6443–6450 (2009). https://doi.org/10.1039/b905802h
17. S.Y. Cho, M. Abdulhafez, M.E. Lee, et al., Promoting Helix-Rich Structure in Silk Fibroin Films through Molecular Interactions with Carbon Nanotubes and Selective Heating for Transparent Biodegradable Devices. ACS Appl Nano Mater acsanm **1**, 5441–5450 (2018). https://doi.org/10.1021/acsanm.8b00784
18. C. Vepari, D.L. Kaplan, Silk as a biomaterial. Prog. Polym. Sci. **32**, 991–1007 (2007). https://doi.org/10.1016/j.progpolymsci.2007.05.013
19. C.Z. Zhou, F. Confalonieri, M. Jacquet, et al., Silk fibroin: Structural implications of a remarkable amino acid sequence. Proteins Struct. Funct. Genet. **44**, 119–122 (2001). https://doi.org/10.1002/prot.1078
20. L.F. Drummy, B.L. Farmer, R.R. Naik, Correlation of the β-sheet crystal size in silk fibers with the protein amino acid sequence. Soft Matter **3**, 877–882 (2007). https://doi.org/10.1039/B701220A
21. J. Guan, D. Porter, F. Vollrath, Thermally induced changes in dynamic mechanical properties of native silks. Biomacromolecules **14**, 930–937 (2013). https://doi.org/10.1021/bm400012k
22. X. Hu, D. Kaplan, P. Cebe, Effect of water on the thermal properties of silk fibroin. Thermochim. Acta **461**, 137–144 (2007). https://doi.org/10.1016/j.tca.2006.12.011
23. J. Pérez-Rigueiro, C. Viney, J. Llorca, M. Elices, Mechanical properties of single-brin silkworm silk. J. Appl. Polym. Sci. **75**, 1270–1277 (2000). https://doi.org/10.1002/(SICI)1097-4628(20000307)75:10<1270::AID-APP8>3.0.CO;2-C
24. C. Viney, J. Llorca, M. Elices, J. Pe, Silkworm silk as an engineering material. J. Appl. Polym. Sci. **70**, 2439–2447 (1998). https://doi.org/10.1002/(SICI)1097-4628(19981219)70:12<2439::AID-APP16>3.0.CO;2-J
25. T.B. Lewis, L.E. Nielsen, Dynamic mechanical properties of particulate-filled composites. J. Appl. Polym. Sci. **14**, 1449–1471 (1970). https://doi.org/10.1002/app.1970.070140604
26. D. Porter, F. Vollrath, K. Tian, et al., A kinetic model for thermal degradation in polymers with specific application to proteins. Polymer (Guildf) **50**, 1814–1818 (2009). https://doi.org/10.1016/j.polymer.2009.01.064
27. D.U. Shah, D. Porter, F. Vollrath, Can silk become an effective reinforcing fibre? A property comparison with flax and glass reinforced composites. Compos. Sci. Technol. **101**, 173–183 (2014). https://doi.org/10.1016/j.compscitech.2014.07.015
28. D.N. Rockwood, R.C. Preda, T. Yücel, et al., Materials fabrication from Bombyx mori silk fibroin. Nat. Protoc. **6**, 1612–1631 (2011). https://doi.org/10.1038/nprot.2011.379
29. L. Meinel, S. Hofmann, V. Karageorgiou, et al., The inflammatory responses to silk films in vitro and in vivo. Biomaterials **26**, 147–155 (2005). https://doi.org/10.1016/j.biomaterials.2004.02.047
30. L.S. Wray, X. Hu, J. Gallego, et al., Effect of processing on silk-based biomaterials: Reproducibility and biocompatibility. J. Biomed. Mater. Res.Part B Appl. Biomater. **99**(B), 89–101 (2011). https://doi.org/10.1002/jbm.b.31875
31. N. Sasithorn, L. Martinová, Fabrication of Silk Nanofibres with Needle and Roller Electrospinning Methods. J. Nanomater. **2014**, 1–9 (2014). https://doi.org/10.1155/2014/947315
32. Y. Tamada, New process to form a silk fibroin porous 3-D structure. Biomacromolecules **6**, 3100–3106 (2005). https://doi.org/10.1021/bm050431f
33. J. Qu, L. Wang, Y. Hu, et al., Preparation of Silk Fibroin Microspheres and Its Cytocompatibility. J. Biomater. Nanobiotechnol. **4**, 84–90 (2013). 10.4028. https://doi.org/10.4236/jbnb.2013.41011
34. U.J. Kim, J. Park, C. Li, et al., Structure and properties of silk hydrogels. Biomacromolecules **5**, 786–792 (2004). https://doi.org/10.1021/bm0345460
35. M.L. Lovett, C.M. Cannizzaro, G. Vunjak-Novakovic, D.L. Kaplan, Gel spinning of silk tubes for tissue engineering. Biomaterials **29**, 4650–4657 (2008). https://doi.org/10.1016/j.biomaterials.2008.08.025

36. L.D. Koh, Y. Cheng, C.P. Teng, et al., Structures, mechanical properties and applications of silk fibroin materials. Prog. Polym. Sci. **46**, 86–110 (2015). https://doi.org/10.1016/j.progpolymsci.2015.02.001
37. C. Fu, Z. Shao, V. Fritz, Animal silks: Their structures, properties and artificial production. Chem. Commun. 6515–6529 (2009). https://doi.org/10.1039/b911049f
38. I.C. Um, H. Kweon, Y.H. Park, S. Hudson, Structural characteristics and properties of the regenerated silk fibroin prepared from formic acid. Int. J. Biol. Macromol. **29**, 91–97 (2001). https://doi.org/10.1016/S0141-8130(01)00159-3
39. L.F. Drummy, D.M. Phillips, M.O. Stone, et al., Thermally induced alpha-helix to beta-sheet transition in regenerated silk fibers and films. Biomacromolecules **6**, 3328–3333 (2005). https://doi.org/10.1021/bm0503524
40. S. Lu, X. Wang, Q. Lu, et al., Insoluble and flexible silk films containing glycerol. Biomacromolecules **11**, 143–150 (2010b). https://doi.org/10.1021/bm900993n
41. Q. Lu, X. Hu, X. Wang, et al., Water-insoluble silk films with silk I structure. Acta Biomater. **6**, 1380–1387 (2010a). https://doi.org/10.1016/j.actbio.2009.10.041
42. O.N. Tretinnikov, Y. Tamada, Influence of casting temperature on the near-surface structure and wettability of cast silk fibroin films. Langmuir **17**, 7406–7413 (2001). https://doi.org/10.1021/la010791y
43. H.J. Jin, J. Park, V. Karageorgiou, et al., Water-stable silk films with reduced β-sheet content. Adv. Funct. Mater. **15**, 1241–1247 (2005). https://doi.org/10.1002/adfm.200400405
44. X. Hu, K. Shmelev, L. Sun, et al., Regulation of silk material structure by temperature-controlled water vapor annealing. Biomacromolecules **12**, 1686–1696 (2011). https://doi.org/10.1021/bm200062a
45. K.H. Kim, L. Jeong, H.N. Park, et al., Biological efficacy of silk fibroin nanofiber membranes for guided bone regeneration. J. Biotechnol. **120**, 327–339 (2005). https://doi.org/10.1016/j.jbiotec.2005.06.033
46. Z. Chen, N. Zhong, J. Wen, et al., Porous Three-Dimensional Silk Fibroin Scaffolds for Tracheal Epithelial Regeneration in Vitro and in Vivo. ACS Biomater. Sci. Eng. **4**, 2977–2985 (2018). https://doi.org/10.1021/acsbiomaterials.8b00419
47. T.K. Mwangi, R.D. Bowles, D.M. Tainter, et al., Synthesis and characterization of silk fibroin microparticles for intra-articular drug delivery. Int. J. Pharm. **485**, 7–14 (2015). https://doi.org/10.1016/j.ijpharm.2015.02.059
48. M. Farokhi, F. Mottaghitalab, Y. Fatahi, et al., Overview of Silk Fibroin Use in Wound Dressings. Trends Biotechnol. **36**, 907–922 (2018). https://doi.org/10.1016/j.tibtech.2018.04.004
49. S. Enomoto, M. Sumi, K. Kajimoto, et al., Long-term patency of small-diameter vascular graft made from fibroin, a silk-based biodegradable material. J. Vasc. Surg. **51**, 155–164 (2009). https://doi.org/10.1016/j.jvs.2009.09.005
50. W. Zhang, L. Chen, J. Chen, et al., Silk Fibroin Biomaterial Shows Safe and Effective Wound Healing in Animal Models and a Randomized Controlled Clinical Trial. Adv. Healthc. Mater. **6**, 1700121 (2017). https://doi.org/10.1002/adhm.20170012
51. Z. Ding, H. Han, Z. Fan, et al., Nanoscale silk-hydroxyapatite hydrogels for injectable bone biomaterials. ACS Appl. Mater. Interfaces **9**, 16913–16921 (2017). https://doi.org/10.1021/acsami.7b03932
52. H. Pan, Y. Zhang, Y. Hang, et al., Significantly reinforced composite Fibers electrospun from silk fibroin/carbon nanotube aqueous solutions. Biomacromolecules **13**, 2859–2867 (2012). https://doi.org/10.1021/bm300877d
53. S. Yodmuang, B.B. Mandal, S.L. McNamara, et al., Silk microfiber-reinforced silk hydrogel composites for functional cartilage tissue repair. Acta Biomater. **11**, 27–36 (2014). https://doi.org/10.1016/j.actbio.2014.09.032
54. C. Wang, X. Li, E. Gao, et al., Carbonized silk fabric for ultrastretchable, highly sensitive, and wearable strain sensors. Adv. Mater. **28**, 6640–6648 (2016). https://doi.org/10.1002/adma.201601572
55. D.A. Frauchiger, A. Tekari, M. Wöltje, et al., A review of the application of reinforced hydrogels and silk as biomaterials for intervertebral disc repair. Eur. Cell. Mater. **34**, 271–290 (2017). https://doi.org/10.22203/eCM.v034a17

56. N. Gogurla, A.K. Sinha, D. Naskar, et al., Metal nanoparticles triggered persistent negative photoconductivity in silk protein hydrogels. Nanoscale **8**, 7695–7703 (2016). https://doi.org/10.1039/C6NR01494A
57. H.J. Jin, S.V. Fridrikh, G.C. Rutledge, D.L. Kaplan, Electrospinning Bombyx mori silk with poly(ethylene oxide). Biomacromolecules **3**, 1233–1239 (2002). https://doi.org/10.1021/bm025581u
58. W.H. Park, L. Jeong, Y.D. Il, S. Hudson, Effect of chitosan on morphology and conformation of electrospun silk fibroin nanofibers. Polymer (Guildf) **45**, 7151–7157 (2004). https://doi.org/10.1016/j.polymer.2004.08.045
59. W. Gong, T. Cheng, Q. Liu, et al., Surgical repair of abdominal wall defect with biomimetic nano/microfibrous hybrid scaffold. Mater. Sci. Eng. C **93**, 828–837 (2018). https://doi.org/10.1016/j.msec.2018.08.053
60. K. Min, M. Umar, H. Seo, et al., Biocompatible, optically transparent, patterned, and flexible electrodes and radio-frequency antennas prepared from silk protein and silver nanowire networks. RSC Adv. **7**, 574–580 (2017). https://doi.org/10.1039/C6RA25580A
61. C. Dionigi, T. Posati, V. Benfenati, et al., A nanostructured conductive bio-composite of silk fibroin–single walled carbon nanotubes. J. Mater. Chem. B **2**, 1424 (2014). https://doi.org/10.1039/c3tb21172j
62. L. Wang, C. Lu, B. Zhang, et al., Fabrication and characterization of flexible silk fibroin films reinforced with graphene oxide for biomedical applications. RSC Adv. **4**, 40312–40320 (2014). https://doi.org/10.1039/c4ra04529g
63. A.J. Mieszawska, N. Fourligas, I. Georgakoudi, et al., Osteoinductive silk-silica composite biomaterials for bone regeneration. Biomaterials **31**, 8902–8910 (2010). https://doi.org/10.1016/j.biomaterials.2010.07.109
64. J. Liu, T. He, G. Fang, et al., Environmentally responsive composite films fabricated using silk nanofibrils and silver nanowires. J. Mater. Chem. C **6**, 12940–12947 (2018). https://doi.org/10.4236/10.1039/C8TC04549F
65. Y. Feng, X. Li, M. Li, et al., Facile preparation of biocompatible silk fibroin/cellulose nanocomposite films with high mechanical performance. ACS Sustain. Chem. Eng. **5**, 6227–6236 (2017). https://doi.org/10.1021/acssuschemeng.7b01161
66. S-W. Hwang, H. Tao, D-H. Kim, et al., A physically transient form of silicon electronics. Science (80-) **337**, 1640–1645 (2012). https://science.sciencemag.org/content/337/6102/1640
67. O. Faruk, A.K. Bledzki, H.P. Fink, M. Sain, Biocomposites reinforced with natural fibers: 2000-2010. Prog. Polym. Sci. **37**, 1552–1596 (2012). https://doi.org/10.1016/j.progpolymsci.2012.04.003
68. A.U. Ude, R.A. Eshkoor, R. Zulkifili, et al., Bombyx mori silk fibre and its composite: A review of contemporary developments. Mater. Des. **57**, 298–305 (2014). https://doi.org/10.1016/j.matdes.2013.12.052
69. D.U. Shah, Developing plant fibre composites for structural applications by optimising composite parameters: A critical review. J. Mater. Sci. **48**, 6083–6107 (2013). https://doi.org/10.1007/s10853-013-7458-7
70. H.Y. Cheung, K.T. Lau, Y.F. Pow, et al., Biodegradation of a silkworm silk/PLA composite. Compos. Part B Eng. **41**, 223–228 (2010). https://doi.org/10.1016/j.compositesb.2009.09.004
71. S.O. Han, H.J. Ahn, D. Cho, Hygrothermal effect on henequen or silk fiber reinforced poly(butylene succinate) biocomposites. Compos. Part B Eng. **41**, 491–497 (2010). https://doi.org/10.1016/j.compositesb.2010.05.003
72. M.P. Ho, K.T. Lau, H. Wang, D. Bhattacharyya, Characteristics of a silk fibre reinforced biodegradable plastic. Compos. Part B Eng. **42**, 117–122 (2011). https://doi.org/10.1016/j.compositesb.2010.10.007
73. M. Takeda, M. Ikeda, S. Satoh, et al., Rab13 is involved in the entry step of hepatitis C virus infection. Acta Med. Okayama **70**, 111–118 (2016). https://doi.org/10.1002/pen
74. Y.Q. Zhao, H.Y. Cheung, K.T. Lau, et al., Silkworm silk/poly(lactic acid) biocomposites: Dynamic mechanical, thermal and biodegradable properties. Polym. Degrad. Stab. **95**, 1978–1987 (2010). https://doi.org/10.1016/j.polymdegradstab.2010.07.015

75. B.B. Mandal, A. Grinberg, E.S. Gil, et al., High-sltrength silk protein scaffolds for bone repair. Proc. Natl. Acad. Sci. U. S. A. **109**, 7699–7704 (2012). https://doi.org/10.1073/pnas.1
76. R. Mi, Y. Liu, X. Chen, Z. Shao, Structure and properties of various hybrids fabricated by silk nanofibrils and nanohydroxyapatite. Nanoscale **8**, 20096–20102 (2016). https://doi.org/10.1039/c6nr07359j
77. S. Ling, C. Li, J. Adamcik, et al., Directed growth of silk nanofibrils on graphene and their hybrid nanocomposites. ACS Macro Lett. **3**, 146–152 (2014). https://doi.org/10.1021/mz400639y
78. Y. Liang, D. Wu, R. Fu, Carbon microfibers with hierarchical porous structure from electrospun fiber-like natural biopolymer. Sci. Rep. **3**, 1–5 (2013). https://doi.org/10.1038/srep01119
79. M. Majibur Rahman Khan, Y. Gotoh, H. Morikawa, M. Miura, Graphitization behavior of iodine-treated Bombyx mori silk fibroin fiber. J. Mater. Sci. **44**, 4235–4240 (2009). https://doi.org/10.1007/s10853-009-3557-x
80. Y.S. Yun, S.Y. Cho, J. Shim, et al., Microporous carbon nanoplates from regenerated silk proteins for supercapacitors. Adv. Mater. **25**, 1993 (2013). https://doi.org/10.1002/adma.201204692
81. C. Wang, K. Xia, M. Jian, et al., Carbonized silk georgette as an ultrasensitive wearable strain sensor for full-range human activity monitoring. J. Mater. Chem. C **5**, 7604–7611 (2017a). https://doi.org/10.1039/c7tc01962a
82. Q. Wang, M. Jian, C. Wang, Y. Zhang, Carbonized silk nanofiber membrane for transparent and sensitive electronic skin. Adv. Funct. Mater. **27**, 1605657 (2017b). https://doi.org/10.1002/adfm.201605657

Chapter 4
Materials Properties and Manufacturing Processes of Nitinol Endovascular Devices

Moataz Elsisy and Youngjae Chun

4.1 Background on Nitinol

4.1.1 History of Nitinol

Shape memory behavior was first discovered in gold–cadmium (AuCd) alloy by Dr. Olander in 1932. The deformed AuCd alloy material in low temperature regains its original shape with the applied heat [1]. Many other alloys, such as CrMn, FePt, $BaTiO_3$, and CuMn, were synthesized to achieve similar phase transformation behavior [2]. By late 1950s at the US Naval Ordnance Laboratory, Buehler and his colleagues discovered the shape memory behavior in nickel–titanium (NiTi) alloy while they worked on intermetallic compounds for heat shielding of missiles [2–4]. Nitinol is referred as NiTi in Ordnance Laboratory and became popular in various fields including aerospace, medical, and other industries due to its inexpensive cost for manufacturing and reliable performance. Nitinol has another unique property, superelasticity, in addition to shape memory behavior. Kurdiumov discovered superelastic behavior of metallic alloys in 1948 by investigating the elastic response of alloys that exhibit phase transformation upon varied stress levels applied on materials [5].

Nitinol has become one of the attractive alloy materials in numerous applications due to two unique properties, shape memory behavior and superelasticity. Currently available commercial products include buckling-resistant antennas, pipe-coupling devices, and eyeglass frames, which utilize superelastic behavior of nitinol [6]. There are more advanced types of applications that use shape memory behavior, which include light structure and engine rotors for aircrafts, biomedical robots, and micro-actuators. More recently, nitinol has been widely used in various medical

M. Elsisy · Y. Chun (✉)
Department of Industrial Engineering, Bioengineering, University of Pittsburgh, Pittsburgh, PA, USA
e-mail: mme41@pitt.edu; yjchun@pitt.edu

P. J. Bártolo, B. Bidanda (eds.), *Bio-Materials and Prototyping Applications in Medicine*, https://doi.org/10.1007/978-3-030-35876-1_4

applications. Dr. Andreasen has developed the first nitinol biomedical application for an orthodontic device utilizing the superelastic behavior of nitinol. Other applications include implantable medical devices, such as endovascular and orthopedic devices [7–9]. Nitinol is specifically beneficial in transcatheter-based devices such as stents, percutaneous heart valves, and vascular occluders, and filters since these devices can be easily collapsed and inserted to a small diameter delivery catheter in low temperature, then deployed to its original shape and dimension in body temperature showing superelastic property after the device delivery.

4.1.2 Macroscopic Behavior of Nitinol

4.1.2.1 Shape Memory Effect

Shape memory effect (SME) is the capability of the material, upon heating, to recover the permanent strain that occurs from the deformation in the martensitic phase [2]. Nitinol has a transformation (T_f) temperature between two different phases, austenite (i.e., above T_f) and martensite (i.e., below T_f). Each phase has different crystal structure that provides the material either to have shape memory effect and superelasticity. The transformation of this alloy could be altered via manipulation of compositional variation of Ni and Ti followed by heat treatment.

When the material is below the transformation temperature (i.e., martensite finish temperature, M_f), it deforms easily with the capability of reaching high level of strains, martensite phase. When the temperature is above the transformation temperature (i.e., austenite finish temperature, A_f), the material exhibits superelastic behavior, austenite phase. Since the temperature is the main parameter for the phase transformation, it is sometimes called thermomechanical transformation. The schematic of phase transformation for the materials is shown in Fig. 4.1. When the material is deformed in martensitic phase, it could recover its deformation through heating to reach the austenite phase as shown in Fig. 4.1a [5].

4.1.2.2 Superelasticity

Elasticity is the capability of recovering from the deformation when the load is removed without generating plastic deformation, similar to the spring effect. Superelastic material is regarded as a very efficient spring. Nitinol compositions can be manipulated in order to attain specific transformation temperatures between phases. Superelasticity happens only to nitinol, when the material temperature lies above the austenite transformation threshold level as shown in Fig. 4.1b.

Nitinol gains relatively high merits over other traditional metals due to the stress-strain response, where nitinol could reach high levels of strains with fracture. Nitinol's superelasticity could exceed the elastic limit of 10%; however, steel and copper fail to precede the elastic limit of 0.5% and 0.1%, respectively [8]. The stress-strain relation of nitinol, compared to steel, is shown in Fig. 4.2a.

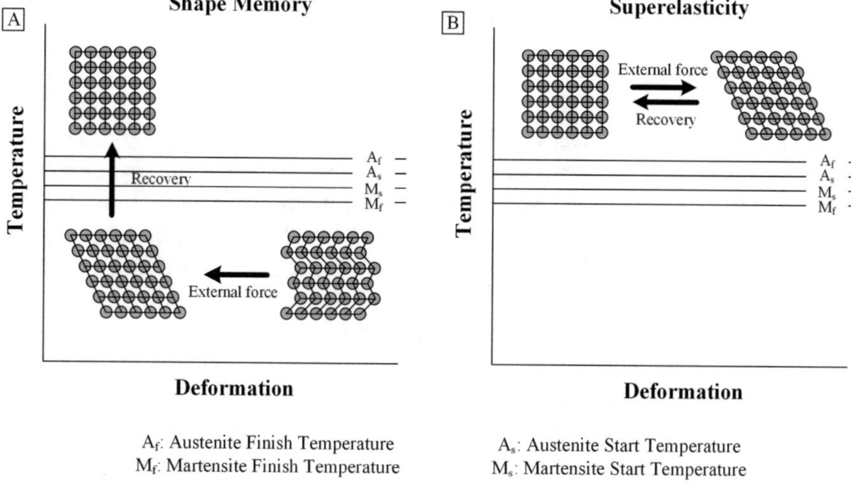

Fig. 4.1 Single crystal model of deformation of nitinol. (**a**) Shape memory effects are exploited by temperature change. The material can recover its shape by heating above austenite finish temperature. (**b**) Superelastic effects only occurs in two-way transformations in the austenite phase

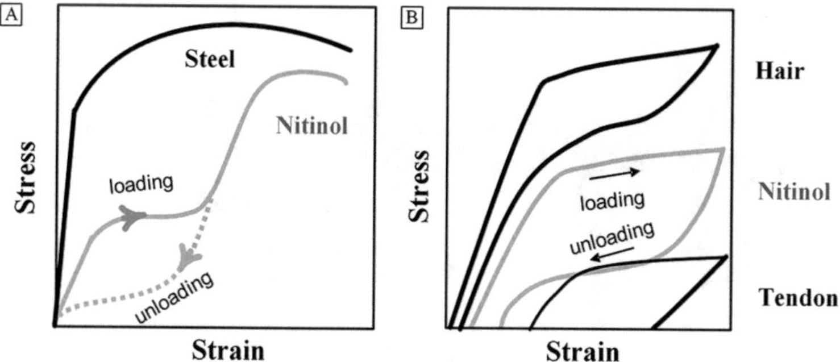

Fig. 4.2 Stress-strain response of nitinol. (**a**) Comparison between nitinol and steel. Nitinol shows hysteresis effect between loading and unloading. (**b**) Loading and unloading of nitinol and living tissues. Nitinol demonstrates similar trend with the living tissues

Another advantage of nitinol for the biomedical application is the exhibition of a similar response trend upon loading and unloading to biological tissues for their deformation as shown in Fig. 4.2b. This phenomenon is called pseudoelasticity, where the material is capable of having a plateau upon large deflections through loading, or recovered deformations through unloading, without noticeable change during loading.

For example, if nitinol is properly manufactured to have its austenite phase transformation temperature less than the body temperature, the material will be fully superelastic in the human body. In this case, nitinol biomedical devices can achieve high strain level, which is a crucial property for medical implants used in the body.

4.1.3 Microscopic View of Nitinol

4.1.3.1 Crystallography

Nitinol has two distinct phases and transforms from one to another by rearranging its crystal structure. The basic structure is composed of nickel and titanium atoms entangled in a single unit cell, which can be cubic, monoclinic, or triclinic structure [10]. The shape recovery of nitinol is considered as a solid–solid phase transformation, also called martensitic transformation. The unit cell in the martensitic transformation changes from cubic (B2: austenite phase) to monoclinic (B19') structure by passing through intermediate phase of orthorhombic (B19).

The martensitic transformation process go through three changes to reach to the final phase as shown in Fig. 4.3 [10]. First, the material is in the cubic structure phase (B2) as the shaded cube shown in Fig. 4.3a, and then the material unit cell transforms into tetragonal shown in Fig. 4.3b. The orthorhombic (C) is formed as an intermediate phase, until homogenous shear force distorted the (B19) phase into the monoclinic (D: B19'). In the final stage, the nickel atom displaces its center position, making the plane lost the center of symmetry [10].

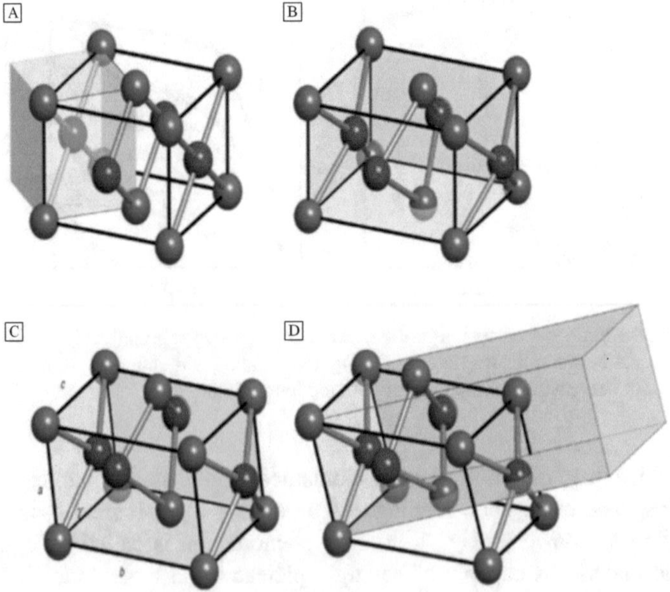

Fig. 4.3 Martensitic distortions of the B2 crystal structure of NiTi. (**a**) The relation between the cubic B2 cell (shaded box) and the undistorted (tetragonal) B19 cell. (**b**) The orthorhombic B19 structure. (**c**) The distortion to the stress stabilized B19' structure. (**d**)The BCO minimum-energy structure with further doubled conventional cell (shaded box). (Reprinted with permission from [10])

The transformation from B2 cubic to B19' monoclinic achieves a large shape change that reaches 7%. This large deformation comes from the crystal asymmetry and produces hysteresis during the transformation phases [11]. While the hysteresis is an important feature to be considered for actuators or springs, it would be less important for the medical implants, specifically for the endovascular device, because superelasticity is a key property for the device, which is not associated with any phase transformation.

4.1.3.2 Martensitic Phase Transformation

The deformation in the martensitic phase transformation occurs without any atom diffusion; however, slip and twinning take place inside the crystal structure. Slip is the motion between slip planes, when the planes have sufficient force to overcome the interatomic forces between slip planes as shown in Fig. 4.4b. Defects in atomic arrangements cause dislocations, responsible for slip motion planes. Slip typically occurs randomly between atomic planes; however, the twinning needs cooperative displacement to happen, which is distinct from the slip, as shown in Fig. 4.4c. Twinning is a rare occurrence in the austenite phase, while it is more common in nitinol martensitic phase, which explains the shape memory behavior of nitinol.

4.1.3.3 Twin Boundary Motion

Twinning (or twin boundary motion) causes change in shape and volume during the martensitic transformation. The twins occur in alternating layers with different directions, which avoid large strains in the martensitic phase. The alternating layers in the twinning boundaries bond with the austenite phase with no twins, which generate an interface of minimum energy planes. The new interface does not produce significant fractures or dislocations as shown in Fig. 4.5 [12].

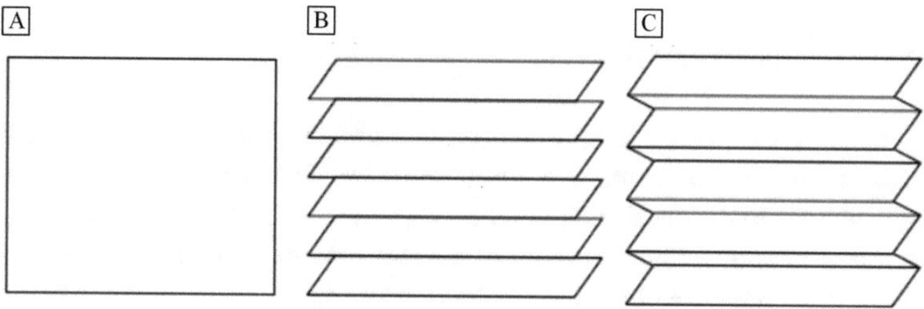

Fig. 4.4 Schematic representation of (**a**) undeformed structure, (**b**) deformed by slip, and (**c**) deformed by twins

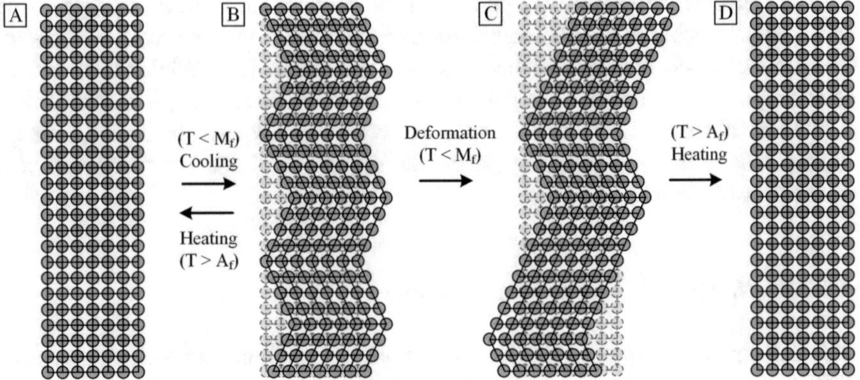

Fig. 4.5 Twinning phenomenon for shape memory mechanism

4.1.4 Other Properties of Nitinol for Endovascular Devices

4.1.4.1 Surface Properties

Biological response of the living organisms to the materials governs the material ability to be implanted in the body. Surface conditions such as roughness and chemical composition have significant influence on the biological response. Higher roughness tends to decrease the hydrophilicity of the surface, which increases protein adsorption (contaminations on the surface) that occurs on the hydrophobic surfaces [13, 14]. Manufacturing processes, such as electropolishing, reduces the roughness of the nitinol surface to minimize the protein adsorption. Also, the chemical composition of the surface can cause inflammatory response due to nickel allergy, upon being implanted inside the body. When nitinol's surface is exposed to the atmosphere or various surface treatments, thin titanium oxide layer is generated, keeping nickel beneath the oxide layer [15]. Most metals including nitinol have high surface energy, which exhibit hydrophilicity. Hydrophilic surface could decrease the protein adsorption during the implantation.

4.1.4.2 Hemocompatibility

Endovascular device's surface should be compatible with blood to minimize thrombosis formation on the device. Nitinol typically shows minimal thrombogenic response, similar to the excellent hemocompatible metallic biomaterials (e.g., stainless steel and titanium) [14, 16]. Nitinol devices that are placed in the blood vessel for a long time have demonstrated better performance of hemocompatibility by minimum protein adsorption, least possible of platelets adhesion, or blood clot formation on the implanted devices. Numerous surface modification strategies including surface coatings, heat treatment, chemical treatment, and electropolishing have been investigated to improve the hemocompatibility in nitinol surface [15, 17].

4.2 Manufacturing Processes of Nitinol Used for Endovascular Devices

Along with the growth of nitinol usage in biomedical devices, manufacturing processes have been developed to produce high-quality nitinol alloy because the composition of nickel and titanium is critical for determining the property of nitinol. The ratio between the two metals is typically 50:50%, and any change in this ratio alters the shape memory and superelastic properties. Any change in nickel or titanium percentage, even by 1%, could shift the transformation temperature with 100 °C [18, 19]. Higher amount of nickel modifies the properties of nitinol to be more superelastic at lower temperatures. Various manufacturing processes and the produced material properties are described in this section, since the manufacturing processes affect the material percentage that determines the nitinol's property.

4.2.1 Drawing Process for Nitinol Wires and Tubes

Continuous production of nitinol wires using the strain annealing ensures the homogeneity of thermomechanical property of the whole product. Figure 4.6 shows the schematic of continuous production of nitinol wires. The main parts are (1) load control unit that regulates the tension during the drawing process, (2) furnace that controls the drawing temperature, and (3) speed control pulley that manipulates the drawing speed. Mechanical properties, final austenite temperature, and straightness are manipulated by temperature, tension, and speed in the drawing process [19]. The drawing is performed under a stress of 35–100 MPa with a temperature range of 450 °C–550 °C.

Nitinol wires undergo several steps through drawing process. Raw material should have 50.5%–51.0% of nickel to produce superelastic wires. Nonmetallic inclusions, micro- and macro-segregations should be avoided before the drawing process [19]. Pre-annealing is performed to form thin oxide layer to work as lubricant through the

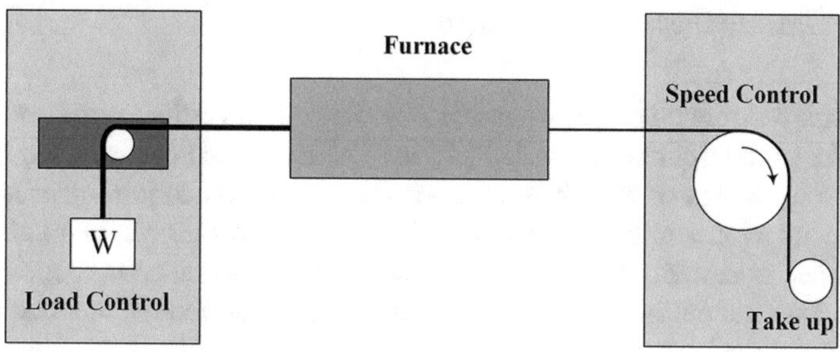

Fig. 4.6 Schematic illustration for drawing of nitinol superelastic wire for continuous production

process; however, if the oxide layer becomes too thick, it will cause cracks in the wire surface [20]. An alternative to the oxide layer as lubricant is molybdenum disulfide, which shows a good lubrication performance in nitinol drawing process [20]. The drawing is performed using multi-passes in inert gas medium through monocrystalline dies. The final pass in drawing is the most important step to produce the desired dimensions and specific superelastic properties. Finally, any lubricant residue is eliminated from the drawn nitinol wire.

4.2.2 Laser-Cutting Process for Stent Fabrication

Laser cutting is a process of applying high-intensity light beam that swiftly heats the targeted area, which melts or/and vaporizes the material through its thickness. Palmaz-Schatz stent was the first approved stent for use in the United States in 1994, which was fabricated using a laser cutting process [21]. Laser cutting offers a couple of advantages over other manufacturing processes, including reliable dimensional accuracy, very small resolution, ease of automation, higher productivity, cut capability for most of materials, and suitable for complex structures. Nitinol laser cutting can be exploited using continuous wave and pulsed wave using different processes including Nd:YAG laser [22], fiber laser [23, 24], and ultrashort pulse laser [25]. The primary laser cutting parameters are cutting speed, laser power, pulse type–duration, and gas (oxygen, inert gas, or air). The main objective of choosing the laser machine and its parameters is to achieve good surface quality with high-dimensional accuracy, as well as minimum heat-affected zone that typically causes brittleness of nitinol surface, in addition to lesser consistent kerf width (distance between cut slot edges). The early laser-cut stents were fabricated using Nd:YAG laser; however, low efficiency and lifetime affect the kerf width consistency [26]. Recently, fiber laser and short-pulsed laser are commonly used for manufacturing the stent due to their reliability, efficiency, and long lifetime.

4.2.3 Joining and Welding Processes for Stents, Filters, Guidewires, and Occluders

Joining and welding processes of nitinol were recently investigated to overcome the drawbacks from laser cutting of the devices. The main problem from the laser cutting is the formation of wide heat-affected zone (relatively to joining process) that alters the microstructure and mechanical properties in nitinol. To minimize the heat-affected zone, wire-to-wire joining was proposed to fabricate the devices. In addition, the laser cutting process is limited by the dimensions of the tube to be cut in order to fabricate device, and laser or other types of welding could be one of the best options for the devices that have dimensions exceed the limits of the fabrication of nitinol tube. Laser-cut process used in stent fabrication typically removes more than 90% of materials, increasing the manufacturing cost, especially for larger devices. Nitinol-to-nitinol laser welding of can be performed using Nd:YAG laser [27, 28], CO_2 laser [29], and tungsten inert gas welding [30]. Thermomechanical properties are usually maintained after the welding process. Nd:YAG laser welding preserves up to 75% of tensile strength of nitinol as well as 7% deformation for the superelastic welded parts [31]. Welding nitinol to dissimilar metal is quite challenging due to the formation of the brittle intermetallic compounds; however, nitinol was successfully welded to stainless steel [28, 32]. There are other types of joining techniques for nitinol such as crimping and swaging.

4.2.4 Subsequent Post-Processes

4.2.4.1 Thermal Annealing

The thermal annealing is performed in order to set the transformation temperature of the nitinol, which is essential to reconstruct the microstructure of the heat affect zone, resulted from the prior fabrication processes. On the one hand, the nitinol material is typically thermally annealed in the temperature around 500 °C to achieve superelastic property in a desired temperature [33]. On the other hand, the nitinol alloys are typically annealed in the temperature between 350 °C and 450 °C for better shape memory behavior. Both thermal annealing temperature and time significantly affect the formation of the oxide layer on the nitinol surface, which govern the thermomechanical and hemocompatible properties of the material [34]. Thermal annealing is also used for shape setting through the relaxation of the material at the desired equilibrium shape. Shape setting could be carried out at temperature around 500 °C using mandrel in order to have the desired geometry of the devices [35].

4.2.4.2 Electropolishing

Electropolishing is the final process for nitinol endovascular devices, which is a standard finishing process for stents, heaver valve fames, or any other implantable devices. This process is typically used for creating smooth surface with corrosion-resistive coating layer on the outer surface of the metal or alloys [36, 37]. The improved surface smoothness is beneficial for endovascular devices due to its enhanced biocompatibility property as implantable devices. Electropolishing reduces the free surface energy to remove any contamination from foreign materials outside the body; thus, thrombosis formation could be minimized [38]. In addition, the electropolishing finishing process has been proven to reduce the nickel concentration on the surface of nitinol alloys [39].

4.3 Nitinol Endovascular Devices

4.3.1 Guidewires

With the growth of endovascular procedures due to their minimally invasive nature, a guidewire is a very important device to deliver the drugs or devices to the location of the diseased or injured blood vessel. Endovascular procedure, sometimes called transcatheter-based procedure, requires a small incision typically in groin to access the vascular system. Due to the nature of less invasive surgery and less postsurgical trauma, endovascular procedures became more popular in the last two decades. A guidewire is typically used with a delivery catheter or sheath that is a kink-resistant composite tube [40, 41]. A guidewire is an ultrathin, elastic, kink-resistant, and sufficiently long metallic wire. A guidewire is used to direct the catheter to the desired location, applying stiffness in the tortuous or bifurcated vessels. The guidewire is first placed in the disease or injury locations in the blood vessel, and then, a catheter is delivered over the guidewire for easy access to the desired locations. Collapsed devices or drugs could be delivered at the location with the catheter (Fig. 4.7).

Figure 4.6a shows a typical structure of the guidewire that consisted of three main parts: central core, tip, and lubricous coating. The central core is the main part that extends through the guidewire, which controls the flexibility, tracking, steering, and support of the guidewire [42]. The central core material is either stainless steel or nitinol. While stainless steel has been extensively used for guidewires previously, however, nitinol is one of the most popular materials used in current guidewire manufacturing because it has superelasticity that is kink-resistant with high torqueability and steering capability [7, 18]. Kinking resistance is specifically very important in guidewire because permanent kinks on the guidewire in endovascular procedure cause difficulties in removing the wire without injuries inside the vascular system [43].

The tip of guidewire is typically covered with very flexible coils, which guide the wire through the lesions without generating any potential complication such as

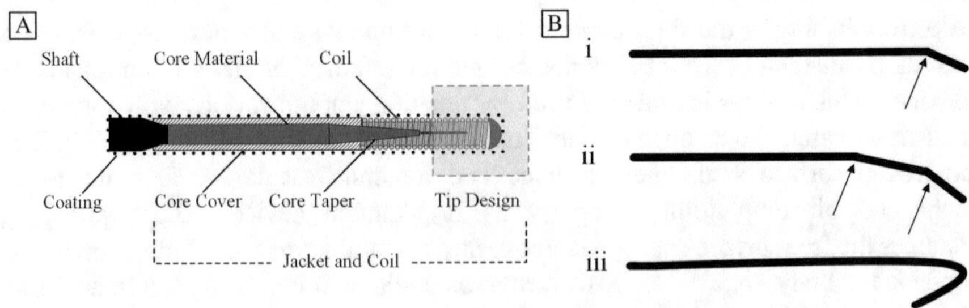

Fig. 4.7 (**a**) Main components of guidewire and (**b**) penetrating tip design the lesion entry; (i) Straight tip or small bend to assist tip penetration, (ii) better navigation and more flexibility are achieved with the secondary bend for a tortuous segment, and (iii) J tips are used to allow returning to the lumen from the subintima

blood vessel perforation or scratch. The tip design governs guidewire's steering performance, flexibility, and pushing capability; therefore, the design is varied depending on the anatomy of the vasculature, e.g., tortuosity, bifurcations, diameter of vessels, and degree of bending needed during navigation. The tip design used for the guidewire depends on the vascular configuration. Figure 4.6b shows various designs used in guidewires; (i) the tip has a small angle bent to increase the penetration, (ii) secondary bend is added to allow better navigation, and (iii) the tip has *"J"* shape to allow returning to the lumen from the subintima. Finally, hydrophilic lubricous coating is applied on the entire guidewire to minimize any potential friction occurred during navigation, to increase physician control, and to provide smooth delivery without generating any thrombogenic issue in the blood stream.

The guidewire performance mainly affects the capability of the radiologists to operate the endovascular operation. Radiologists were requested to use three commercial guidewires and evaluate them in terms of specific parameters. The performance parameters were torque response, tortuous vessels navigation, radiopacity (ability to be seen in X-rays), balance, lubricity, and tip shape retention. The guidewires used in this survey are ZIPwire™ Hydrophilic Guide Wire, HiWire® Nitinol Core Wire Guide, and TERUMO GLIDEWIRE®. The evaluation is represented in Fig. 4.8 (data from [44]).

4.3.2 Self-Expanding Stents

Stent is a mesh scaffold used to widen the narrowed blood vessel due to atherosclerosis or plaque formation, typically inside the coronary, intracranial, or peripheral arteries [45–47]. The stents are also used for blocking dilated artery (i.e., aneurysm) or for inserting a new conduit in the aneurysm locations in order to prevent potential

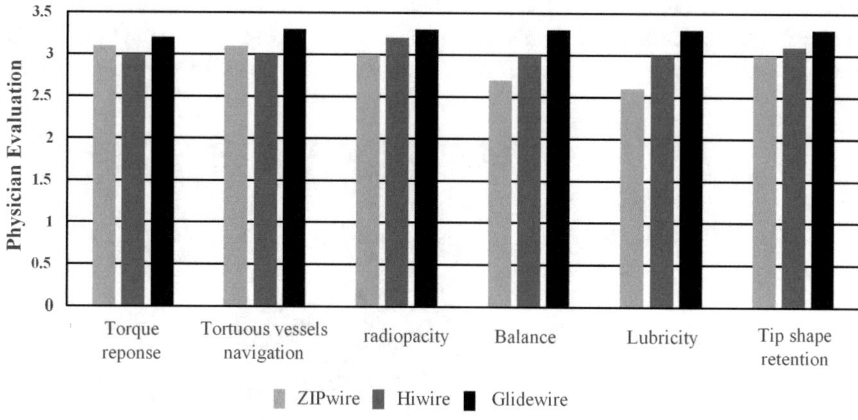

Fig. 4.8 The rating of three representative commercial hydrophilic-coated guidewire characteristics, evaluated by 10 physicians in 40 different cases for every wire. (Data from [44])

rupture of aneurysms. Intravascular stenting technique was first investigated by Charles Dotter in 1969 [48]. This technique was developed to reconstruct vascular patency using transcatheter procedure, in a way other than percutaneous balloon angioplasty, which temporarily dilates the narrowed blood vessel with the inflated balloon to break the plaques or to push away the fatty substances to the wall [49]. Most of the devices in this era were coil spring-shaped stents. More than one decade later, Julio Palmaz developed a tube-shaped balloon expandable stainless steel stent for small artery applications [21].

Balloon inflation is needed to deploy the stainless steel stent to generate permanent plastic deformation in stent because stainless steel is ductile in the range of expansion from the delivery catheter size to the target artery size [50, 51]. Although balloon expandable stents have been widely used in earlier stent history, there are potential vessel wall damages that lead to severe restenosis (i.e., tissue ingrowth) due to the overexpanded balloon dilation to place the stent [52]. Therefore, self-expandable stents were widely investigated to replace the balloon expandable stents, which is simple to deploy, generating minimal intima damages [53]. One of the best metallic materials for self-expanding stents is nitinol as described in the earlier section (Sect. 1.2). There are primarily two types of nitinol stents: (1) laser-cut stent from thin wall nitinol tube for small diameter applications (e.g., coronary artery, neurovascular, or peripheral arterial stents) and (2) wire braided or bent stent for large diameter applications (e.g., thoracic or abdominal aortic aneurysm stent grafts). Figure 4.9 shows representative commercially available nitinol stent products for peripheral artery treatments. Figure 4.9a represents the Innova™ (Boston Scientific) stent, which can be used for proximal popliteal artery and superficial femoral artery. Figure 4.9b shows VIABAHN® (Gore Medical) stent graft, where the metallic backbone is covered by thin polymer, typically either expanded polytetrafluoroethylene (ePTFE) or Dacron polyester. This type of stent is used to graft the disease tissue and also to protect aortic–thoracic aneurysms from rupture [54, 55].

In addition, nitinol stent is non-ferromagnetic, which is MRI compatible. Patients who have stents in the body do not have problems for taking MRI in the future after the placement of the stents [56]. The common approach used for the stent delivery

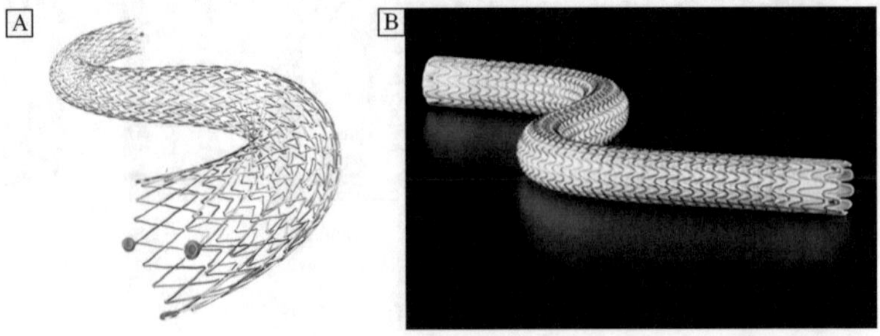

Fig. 4.9 (**a**) Innova™ Stent Delivery System (Boston Scientific). (**b**) VIABAHN® Endoprosthesis (Gore Medical)

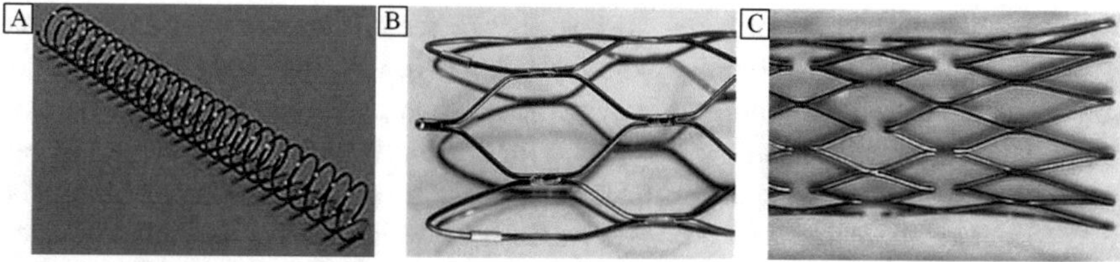

Fig. 4.10 (**a**) Intracoil stent (IntraTherapeutics). (**b**) Wire-based stents connected using welding, symphony stent (Boston Scientific). (**c**) Laser-cut nitinol tubular stents, Memotherm (Bard Angiomed). (All images are reprinted with the permission from [8])

is keeping the stents in the cooler martensitic phase for crimping and then inserting into the catheter. Once the catheter reaches to the desired location inside the body, it is removed to deploy the stent with its nitinol's own superelasticity.

Earlier stents were manufactured by transforming the nitinol wires into a coil as shown in Fig. 4.10a; however, due to low radial stiffness and collapsibility, these designs were abandoned. Current nitinol stents are wire-, sheet-, or tube-based designs. Wire-based stents are made of round or flat wires, which can be connected via welding or any other joining methods as shown in Fig. 4.10b. One advantage of wire-based stents is the ability to easily retrieve the stent in certain applications because the stent struts are typically closed-cell geometry (i.e., all wires are connected) [8]. Sheet-based stent is used to overcome the wire-based stent's high collapsed volume due to the multiple crossed wires. The sheet is first laser cut, and then rolled up and welded to make the circular profile. While the sheet-based stent was preferred for a smaller stent, there were relatively bulky joining regions that should be avoided for better collapsing with a small catheter. Tube-based stents are seamless stents (Fig. 4.10c), manufactured by laser cutting from the nitinol tube, to avoid any potential fracture issue that may occur in the sheet-based stent due to the welding regions. The stents are annealed after the manufacturing processes to set the transformation temperature for the desired thermomechanical properties. Additionally, this thermal process eliminates any stress concentration zone that limits the localized brittleness. Radiopaque makers are added on both ends or any desired locations within the stent to help position the stent under fluoroscopic guidance.

4.3.3 Percutaneous Heart Valve Frame

Symptomatic aortic stenosis (AS) is considered one of the main valvular heart disease, which is an important source of cardiovascular morbidity and death worldwide [57]. Aortic stenosis is the narrowing of the valve that delivers high pressure oxygenated blood from the heart to the aortic artery, which carries the blood to the main abdominal organs and the lower part of the body. The standard treatment of aortic valve stenosis is open chest aortic valve replacement; however, most of the patients are elderly who may not survive due to the open-surgery complications. Thus, one-third of the patients are rejected for surgery [58]. Less invasive techniques were adapted to overcome the open-chest complications. Percutaneous aortic valve replacement has developed as a new promising technique in the recent years as minimum invasive operation for symptomatic (AS) treatment [59, 60]. The first percutaneous valve was proposed in the early 1990s using balloon expanded valve stent [61]. With the aid of superelasticity in nitinol alloy (i.e., self-expanding device) nitinol-based aortic valve was used for the minimally invasive treatment of aortic stenosis. The metallic backbone is manufactured using laser cutting in a similar way of stent. Figure 4.11a shows a representative commercially available nitinol heart valve which is CoreValve Revalving™ System that is integrated with porcine pericardial trileaflet valve sewn to the nitinol backbone using ePTFE sutures [62]. The placement of the valve is performed under fluoroscopic guidance as shown in Fig. 4.11b. Figure 4.11c shows the implanted heart valve after a few months, which shows that the nitinol valve replaced the aortic valve well.

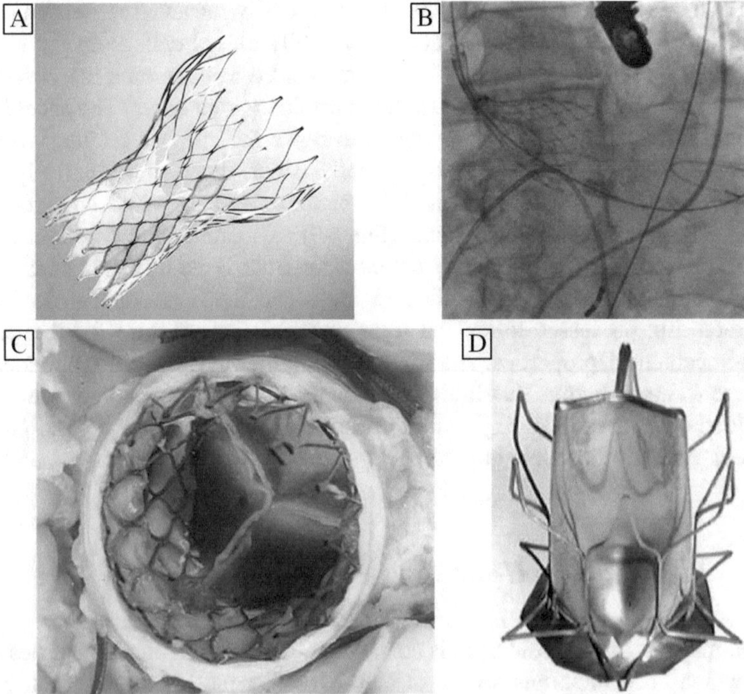

Fig. 4.11 (**a**) CoreValve Revalving System profile (reprint with the permission from [57]). (**b**) In situ Coronal MRI image CoreValve prosthesis, after six months of implantation. (Reprint with the permission from [58]). (**c**)Fluoroscopic guidance for implanting the CoreValve prosthesis. (Reprint with the permission from [58])

4.3.4 Atrial Septal Defect (ASD) Occluder

Atrial septal defect (ASD) is a hole that occurs in the septum, the wall that separates between the upper chambers of the heart. This defect allows the oxygenated blood, comes from the lungs in the left atria, to induce a leakage to the poor-oxygenated blood in the right atria. Figure 4.12 is the schematic of the atrial septal defect. The patients with this defect suffer from fatigue, arrhythmias, and congestive blood failure [63]. ASD is a common congenital heart defect with an occurrence rate of 3.78/10000 live births, considered the fourth frequent form of congenital heart disease [64]. Even though open-heart surgical operation is widely accepted to repair the defect, exploiting minimally invasive alternative has been recently developed to overcome any complication found in the open-heart surgery [65].

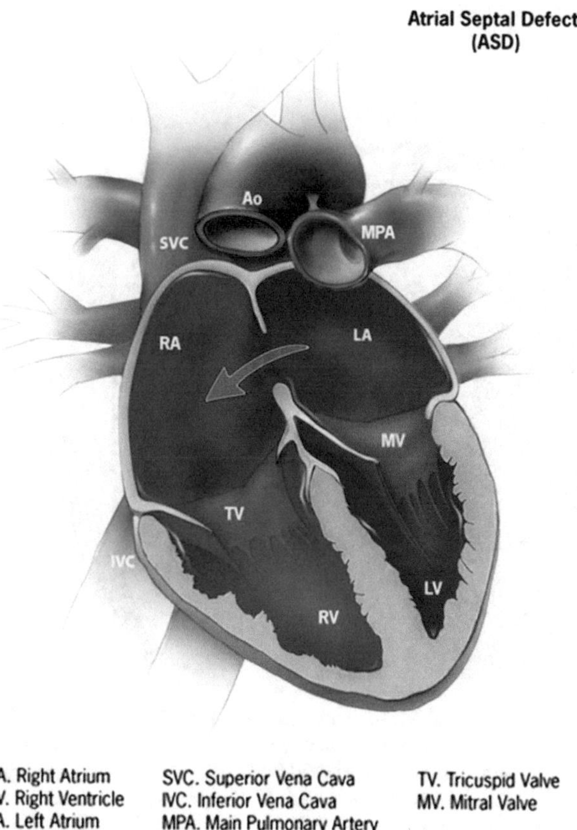

Fig. 4.12 Schematic of arterial septal defect in the heart

King and Mills did the first endovascular occluder trial to repair ASD during cardiac catheterization in 1974. They used the opposed pair of stainless steel umbrellas covered with Dracon [63]. This device had many problems including structural failure and inability to recapture [66]. With the significant development of shape memory alloys, nitinol was used to repair the atrial septal defect with the self-expanding, self-centered, and repositionable properties.

The ASD occluder was named the Amplatzer, which consists of two round disks of nitinol mesh. The mesh was made from 0.004- to 0.005-inch nitinol wire, which is tightly woven into two flat buttons, and the disks were linked using connecting waist [66], as shown in Fig. 4.13a. The whole device can be delivered via either 6Fr or 7Fr sheath (Fig. 4.13b), depending on the diameter wire and expanded size of the occluders [67]. The two flat retention meshes extend to the radial direction beyond the central waist (Fig. 4.13c) to work as secure anchors. The nitinol mesh is covered with Dacron. The Amplatzer is deployed with ultrasound and fluoroscopic guide to ensure optimal positioning [65]. Figure 4.13d shows cineradiographic frames of the ASD occluder implanted inside human body. The occluder demonstrates good positioning and whole closure.

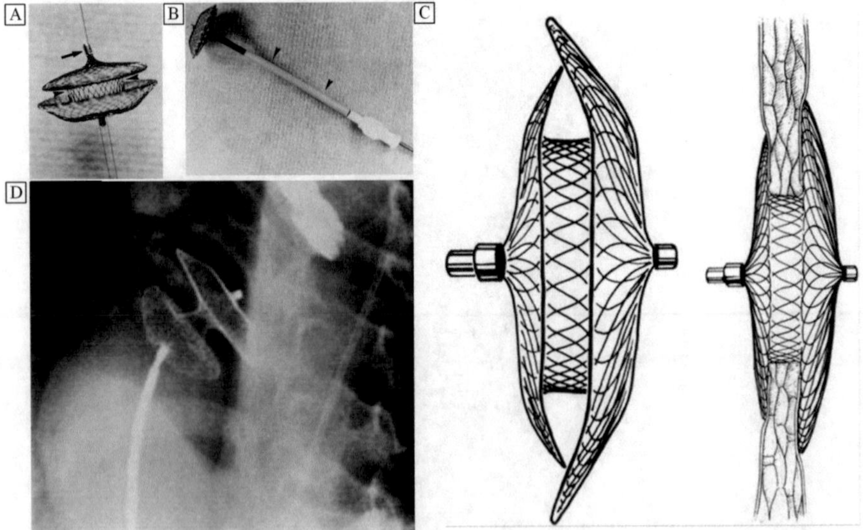

Fig. 4.13 (**a**) Amplatzer septal occluder made of two round disks from 0.005-in. nitinol wire, that was tightly woven into with a 4-mm connecting waist (arrowheads). (Reprint with the permission from [66]). (**b**) Adaptor tube (arrowheads) used for the occluder delivery. (Reprint with the permission from [66]). (**c**) Mechanisms of leakage-proof closure with the two retention disks, which are angled inward. The left atrial retention disk is slightly bigger than the right, to ensure overfitted clamping against around the defect. (Reprint with the permission from [67]). (**d**) Good positioning and complete closure of the Amplatzer are shown using Levo-phase of pulmonary arteriogram. (Reprint with the permission from [66])

4.3.5 Vena Cava Filter

The vena cava is a large vein that carries the deoxygenated blood to the heart. The human body has two venae cavae: the superior vena cava that carries the blood from head and upper body and the inferior vena cava that carries blood from the lower body. Clot formed in these veins could travel to various locations and may cause fatal complications in the brain or lungs with the clots or fragmentation from clots (i.e., clot embolization) [68]. Due to the device's large deployed diameter, it is always challenging to deploy endovascular devices [69]. However, these filter devices can be easily collapsed in low temperature region (i.e., martensite phase) by cooling, then, the filters recovers its original shape in the higher temperature (i.e., austenite phase) such as human body temperature, which is called shape memory effect of nitinol material. The first nitinol filter was presented by M. Simon in 1977. Shape memory effect of nitinol was used to make the filter recover its predefined configuration, when the filter is heated by body temperature upon implanting. Figure 4.14a shows the filter inside the body, with full recovery and good positioning [70]. The test results for vena cava demonstrate the capability of IVC filter to collect the blood clots [68, 71, 72]. More recently, new filter devices have been introduced. Figure 4.14b shows the commercially available Optease vena cava filter (Cordis), which is used as a temporary filter. The filter consists of six diamond-shaped laser-cut nitinol struts. The filter contains a self-centering portion and the upper hook for easy retrieval of the filter. Another commercial nitinol filter is Denali vena cava filter (Bard, Fig. 4.14c). The filter consists of 12 laser-cut nitinol appendages, which can be used for temporary or permanent placement of the device.

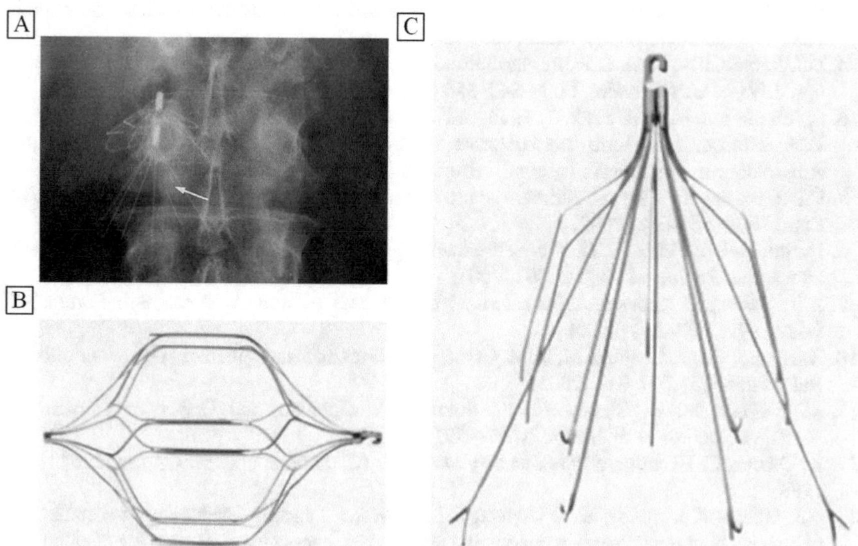

Fig. 4.14 (**a**) Abdominal plain film. The Simon nitinol filter is well positioned in the inferior vena cava. (Reprint with the permission from [70]). (**b**) Optease vena cava filter (Cordis). (Reprint with the permission of [73]). (**c**) Denali vena cava filter (Bard). (Reprint with the permission of [73])

4.4 Summary

Many endovascular devices, which are commercially available and under research, take advantage of the mechanical and biological properties of nitinol for the treatment of various physiological problems. Both shape memory alloy and superelastic properties are utilized to achieve the capability of delivery of the endovascular devices with minimum invasive techniques, and self-expanding performance. Manufacturing processes influence thermomechanical behaviors of nitinol, hence should be carefully selected for various devices. Nitinol alloy have been applied for various endovascular devices, including guidewires, stents, percutaneous heart valves, filters, and occluders, demonstrating excellent performance in terms of hemocompatibility, surface properties, and mechanical reliability. Many nitinol endovascular devices are currently under research; therefore, these could be used in the future for better clinical outcomes, helping patients who have vascular diseases or injuries.

References

1. A. Ölander, An electrochemical investigation of solid cadmium-gold alloys. J. Am. Chem. Soc. **54**(10), 3819–3833 (1932)
2. C.M. Wayman, J.D. Harrison, W.J. Bryan, Shape memory effect. JOM **41**, 26–28 (1989)
3. W.J. Buehler, J.V. Gilfrich, R.C. Wiley, Effect of low-temperature phase changes on the mechanical properties of alloys near composition TiNi. J. Appl. Phys. **34**(5), 1475–1477 (1963)
4. F.E. Wang, W.J. Buehler, S.J. Pickart, Crystal structure and a unique "martensitic" transition of TiNi. J. Appl. Phys. **36**(10), 3232–3239 (1965)
5. D.J. Rabkin, E.V. Lang, D.P. Brophy, Nitinol properties affecting uses in interventional radiology. J. Vasc. Interv. Radiol. **11**(3), 343–350 (2000)
6. S. Yamamoto, H.E. Wilczek, T. Iwata, M. Larsson, H. Gjertsen, G. Söderdahl, G. Solders, B.G. Ericzon, Long-term consequences of domino liver transplantation using familial Amyloidotic polyneuropathy grafts. Transpl. Int. **20**(11), 926–933 (2007)
7. C.D.J. Barras, K.A. Myers, Nitinol – its use in vascular surgery and other applications. EJVES Extra **19**(6), 564–569 (2000)
8. D. Stoeckel, A. Pelton, T. Duerig, Self-expanding nitinol stents: Material and design considerations. Eur. Radiol. **14**(2), 292–301 (2004)
9. A.R. Pelton, D. Stöckel, T.W. Duerig, Medical uses of nitinol. Mater. Sci. Forum **328** (May 1999), 327–328 (2000)
10. X. Huang, G.J. Ackland, K.M. Rabe, Crystal structures and shape-memory behaviour of NiTi. Nat. Mater. **2**(5), 307–311 (2003)
11. M. Meyers and K. Chawla, "Solid Solution, Preciptation, and Dispersion Hardening," *Mechanical Behavior of Materials*, 558–591 (2009)
12. K. Otsuka, C. Wayman, *Shape Memory Materials* (Cambridge university press, Cambridge, 1998)
13. N.J. Hallab, K.J. Bundy, K. O'Connor, R.L. Moses, J.J. Jacobs, Evaluation of metallic and polymeric biomaterial surface energy and surface roughness characteristics for directed cell adhesion. Tissue Eng. **7**(1), 55–71 (2001)

14. L. Ponsonnet, V. Comte, A. Othmane, C. Lagneau, M. Charbonnier, M. Lissac, N. Jaffrezic, Effect of surface topography and chemistry on adhesion, orientation and growth of fibroblasts on nickel-titanium substrates. Mater. Sci. Eng. C **21**(1–2), 157–165 (2002)
15. S.A. Shabalovskaya, Surface, corrosion and biocompatibility aspects of nitinol as an implant material. Biomed. Mater. Eng. **12**(1), 69–109 (2002)
16. D.A. Armitage, T.L. Parker, D.M. Grant, Biocompatibility and hemocompatibility of surface-modified NiTi alloys. J. Biomed. Mater. Res. Part A **66**(1), 129–137 (2003)
17. S. Sapatnekar, K.M. Kieswetter, K. Merritt, J.M. Anderson, L. Cahalan, M. Verhoeven, M. Hendriks, B. Fouache, P. Cahalan, Blood–biomaterial interactions in a flow system in the presence of bacteria: Effect of protein adsorption. J. Biomed. Mater. Res. **29**(2), 247–256 (1995)
18. T. W. Duerig, K. Melton, and D. Stöckel, Engineering aspects of shape memory alloys: Butterworth-Heinemann (2013)
19. L. M. Schetky and M. Wu, "Issues in the further development of Nitinol properties and processing for medical device applications," in Medical device materials: proceedings from the materials & processes for medical devices conference. 271–276 (2003, 2004)
20. S. Shabalovskaya, J. Anderegg, J. Van Humbeeck, Critical overview of Nitinol surfaces and their modifications for medical applications. Acta. Biomater. **4**, 447–467 (2008)
21. J. Palmaz, Progress in radiology balloon-expandable. Am. J. Roentgenol. **150**, 1263–1269 (1988)
22. R. Pfeifer, D. Herzog, M. Hustedt, S. Barcikowski, Pulsed Nd : YAG laser cutting of NiTi shape memory alloys — Influence of process parameters. J. Mater. Process. Tech. **210**(14), 1918–1925 (2010)
23. C.A. Biffi, A. Tuissi, Nitinol laser cutting: Microstructure and functional properties of femtosecond and continuous wave laser processing. Smart Mater. Struct. **26**, 035006 (2017)
24. L. Liu, D. Bo, L. Yi, F. Tong, Y. Fu, Fiber laser micromachining of thin NiTi tubes for shape memory vascular stents. Appl. Phys. A Mater. Sci. Process. **122**(7), 1–9 (2016)
25. N.M.D.W.A. Boor, W.O.Z.L.L. Li, Picosecond laser micromachining of nitinol and platinum – Iridium alloy for coronary stent applications. Appl. Phys. A Mater. Sci. Process., 607–617 (2012)
26. L. Chen, L. Chen, Laser cutting for medical device (stent) – Yesterday, today and tomorrow, ICALEO® 2008 Congr. Proc., 607 (2008)
27. L. Quintino, R.M. Miranda, U.N. De Lisboa, Welding shape memory alloys with NdYAG lasers. Soldag. e Inspecção **17**, 210–217 (2012)
28. G.R. Mirshekari, A. Saatchi, A. Kermanpur, S.K. Sadrnezhaad, Laser welding of NiTi shape memory alloy: Comparison of the similar and dissimilar joints to AISI 304 stainless steel. Opt. Laser Technol. **54**, 151–158 (2013)
29. Y.T. Hsu, Y.R. Wang, S.K. Wu, C. Chen, Effect of CO2 laser welding on the shape-memory and corrosion characteristics of TiNi alloys. Metall. Mater. Trans. A **32**(March), 569–576 (2001)
30. J.P. Oliveira, D. Barbosa, F.M.B. Fernandes, R.M. Miranda, Tungsten inert gas (TIG) welding of Ni-rich NiTi plates: Functional behavior. Smart Mater. Struct. **25**(3), 03LT01 (2016)
31. M.H. Wu, Fabrication of nitinol materials and components. Mater. Sci. Forum **395**, 285–292 (2002)
32. H.M. Li, D.Q. Sun, X.L. Cai, P. Dong, W.Q. Wang, Laser welding of TiNi shape memory alloy and stainless steel using Ni interlayer. Mater. Des. **39**, 285–293 (2012)
33. M. Drexel, G. Selvaduray, A. Pelton, The effects of cold work and heat treatment on the properties of nitinol wire. Med. Device Mater. IV Proc. Mater. Process. Med. Devices Conf. **2007**, 114–119 (2008)
34. L. Xu, R. Wang, Y. Liu, The optimization of annealing and cold-drawing in the manufacture of the Ni-Ti shape memory alloy ultra-thin wire. Int. J. Adv. Manuf. Technol. **55**(9–12), 905–910 (2011)

35. Vojtěch, D., Influence of heat treatment of shape memory NiTi alloy on its mechanical properties, Proceeding 19th conference Metall. Mater. "Metal 2010," 2010, pp. 867–871
36. W. Simka, M. Kaczmarek, A. Baron-Wiecheć, G. Nawrat, J. Marciniak, J. Zak, Electropolishing and passivation of NiTi shape memory alloy. Electrochim. Acta **55**(7), 2437–2441 (2010)
37. R. Venugopalan, C. Trépanier, Assessing the corrosion behaviour of nitinol for minimally-invasive device design. Minim. Invasive Ther. Allied Technol. **9**(2), 67–73 (2000)
38. B. Thierry, M. Tabrizian, O. Savadogo, L. Yahia, Effects of sterilization processes on NiTi alloy: Surface characterization. J. Biomed. Mater. Res. **49**(1), 88–98 (2000)
39. A. Michiardi, C. Aparicio, J.A. Planell, F.J. Gil, New oxidation treatment of NiTi shape memory alloys to obtain Ni-free surfaces and to improve biocompatibility. J. Biomed. Mater. Res. Part B Appl. Biomater. **77**(2), 249–256 (2006)
40. Dankelman, J., Grimbergen, C. A., and Stassen, H. G., 2004, Engineering for Patient Safety
41. L. Aklog, D.H. Adams, G.S. Couper, R. Gobezie, S. Sears, L.H. Cohn, A.F. Carpentier, D.B. Skinner, Techniques and results of direct-access minimally invasive mitral valve surgery: A paradigm for the future. J. Thorac. Cardiovasc. Surg. **116**(5), 705–715 (1998)
42. C. Walker, "Guidewire selection for peripheral vascular interventions," Endovasc. Today **5**, 80–83 (2013)
43. T. Anson, Shape memory alloys - medical applications. Mater. World **212**, 745–747 (2010)
44. A. Shah, C. Lau, S.W. Stavropoulos, A. Nemeth, M.C. Soulen, J.A. Solomon, J.I. Mondschein, A.A. Patel, R.D. Shlansky-goldberg, M. Itkin, J.L. Chittams, S.O. Trerotola, Comparison of physician-rated performance characteristics of hydrophilic-coated guide wires. J. Vasc. Interv. Radiol. **19**, 400–405 (2008)
45. Y. Chun, D.S. Levi, K.P. Mohanchandra, F. Vinuela, F. Vinuela, G.P. Carman, Thin film nitinol microstent for aneurysm occlusion. J. Biomech. Eng. **131**(5), 051014 (2009)
46. R. Uflacker, J. Robison, Endovascular treatment of abdominal aortic aneurysms : A review. Eur. Radiol. **11**, 739–753 (2001)
47. H.A. Gary, A.W. Crane, J.A. Kaufman, S.C. Geller, D.C. Brewster, C. Fan, R.P. Cambria, G.M. Lamuraglia, J.P. Gertler, W.M. Abbott, A.C. Waltman, A.W. Crane, Endovascular repair of abdominal aortic aneurysms: Current status and future directions. Am. J. Roentgenol. **175**, 289–302 (2000)
48. C.T. Dotter, Transluminally-placed coilspring endarterial tube grafts. Long-term patency in canine popliteal artery. Investig. Radiol. **4**(5), 329–332 (1969)
49. J.C. Palmaz, Intravascular stenting: From basic research to clinical application. Cardiovasc. Intervent. Radiol. **15**(5), 279–284 (1992)
50. R.A. Schatz, J.C. Palmaz, F.O. Tio, F. Garcia, O. Garcia, S.R. Reuter, Balloon-expandable intracoronary stents in the adult dog. Circulation **76**(2), 450–457 (1987)
51. A. Buchwald, C. Unterberg, G. Werner, H. Kreuzer, V. Wiegand, E. Voth, Initial clinical results with the Wiktor stent: A new balloon-expandable coronary stent. Clin. Cardiol. **14**(5), 374–380 (1991)
52. W. Kurre, F. Brassel, R. Brüning, J. Buhk, B. Eckert, S. Horner, M. Knauth, T. Liebig, J. Maskova, D. Mucha, V. Sychra, M. Sitzer, M. Sonnberger, M. Tietke, J. Trenkler, B. Turowski, J. Berkefeld, Complication rates using balloon-expandable and self-expanding stents for the treatment of intracranial atherosclerotic stenoses. Neuroradiology **54**(1), 43–50 (2012)
53. R. Beyar, R. Shofti, E. Grenedier, M. Henry, O. Globerman, M. Beyar, Self-expandable nitinol stent for cardiovascular applications: Canine and human experience. Catheter. Cardiovasc. Diagn. **32**(2), 162–170 (1994)
54. R.K. Greenberg, K. West, K. Pfaff, J. Foster, D. Skender, S. Haulon, J. Sereika, L. Geiger, S.P. Lyden, D. Clair, L. Svensson, B. Lytle, Beyond the aortic bifurcation: Branched endovascular grafts for thoracoabdominal and aortoiliac aneurysms. J. Vasc. Surg. **45**(5), 879–886 (2006)
55. M. Grabenwöger, D. Hutschala, M.P. Ehrlich, F. Cartes-Zumelzu, S. Thurnher, J. Lammer, E. Wolner, M. Havel, Thoracic aortic aneurysms: Treatment with endovascular self-expandable stent grafts. Ann. Thorac. Surg. **69**(2), 441–445 (2000)

56. T. Duerig, A. Pelton, D. Sto, An overview of nitinol medical applications. Mater. Sci. Eng. A **275**, 149–160 (1999)
57. A. Zajarias, A.G. Cribier, Outcomes and safety of percutaneous aortic valve replacement. J. Am. Coll. Cardiol. **53**(20), 1829–1836 (2009)
58. J. Baan, Z.Y. Yong, K.T. Koch, J.P. Henriques, B.J. Bouma, S.G. de Hert, J. van der Meulen, J.G. Tijssen, J.J. Piek, B.A. de Mol, Percutaneous implantation of the CoreValve aortic valve prosthesis in patients at high risk or rejected for surgical valve replacement: Clinical evaluation and feasibility of the procedure in the first 30 patients in the AMC-UvA. Neth. Heart J. **18**(1), 18–24 (2010)
59. E. Grube, J.C. Laborde, B. Zickmann, U. Gerckens, T. Felderhoff, B. Sauren, A. Bootsveld, L. Buellesfeld, S. Iversen, First report on a human percutaneous transluminal implantation of a self-expanding valve prothesis for interventional treatment of aortic valve stenosis. Catheter. Cardiovasc. Interv. **66**(4), 465–469 (2005)
60. Y. Boudjemline, P. Bonhoeffer, Steps toward percutaneous aortic valve replacement. Circulation **105**(6), 775–778 (2002)
61. H.R. Andersen, L.L. Knudsen, J.M. Hasenkam, Transluminal implantation of artificial heart valves. Description of a new expandable aortic valve and initial results with implantation by catheter technique in closed chest pigs. Eur. Heart J. **13**(5), 704–708 (1992)
62. M.B. Leon, S. Kodali, M. Williams, M. Oz, C. Smith, A. Stewart, A. Schwartz, M. Collins, J.W. Moses, Transcatheter aortic valve replacement in patients with critical aortic stenosis: Rationale, device descriptions, early clinical experiences, and perspectives. Semin. Thorac. Cardiovasc. Surg. **18**(2), 165–174 (2006)
63. T.D. King, S.L. Thompson, C. Steiner, N.L. Mills, Secundum atrial septal defect: Nonoperative closure during cardiac catheterization. JAMA J. Am. Med. Assoc. **235**(23), 2506–2509 (1976)
64. G.C. Emmanoulides, H.D. Allen, R. T, *Heart Disease in Infants, Children and Adolescents, Including the Fetus and Young Adults* (Williams and Wilkins, Baltimore, 1995)
65. J.G. Murphy, B.J. Gersh, M.D. McGoon, D.D. Mair, C.J. Porter, D.M. Ilstrup, D.C. McGoon, F.J. Puga, J.W. Kirklin, G.K. Danielson, Long-term outcome after surgical repair of isolated atrial septal defect. N. Engl. J. Med. **323**(24), 1645–1650 (1990)
66. B.D. Thanopoulos, C.V. Laskari, G.S. Tsaousis, A. Zarayelyan, A. Vekiou, G.S. Papadopoulos, Closure of atrial septal defects with the Amplatzer occlusion device: Preliminary results. J. Am. Coll. Cardiol. **31**(5), 1110–1116 (1998)
67. M.J.A. Sharafuddin, X. Gu, J.L. Titus, M. Urness, J.J. Cervera-Ceballos, K. Amplatz, Transvenous closure of Secundum atrial septal defects: Preliminary results with a new self-expanding nitinol prosthesis in a swine model. Circulation **95**(8), 2162–2168 (1997)
68. M. Simon, R. Kaplow, E. Salzman, D. Freiman, A vena cava filter using thermal shape memory alloy experimental aspects. Radiology **125**(1), 89–94 (1977)
69. E. Bruckheimer, A.G. Judelman, S.D. Bruckheimer, I. Tavori, G. Naor, B.T. Katzen, In vitro evaluation of a retrievable low-profile nitinol vena cava filter. J. Vasc. Interv. Radiol. **14**(4), 469–474 (2003)
70. P.A. Poletti, C.D. Becker, L. Prina, P. Ruijs, H. Bounameaux, D. Didier, P.A. Schneider, F. Terrier, Long-term results of the Simon nitinol inferior vena cava filter. Eur. Radiol. **8**(2), 289–294 (1998)
71. E. Engmann, M.R. Asch, Clinical experience with the antecubital Simon nitinol IVC filter. J. Vasc. Interv. Radiol. **9**(5), 774–778 (1998)
72. B. FThomas, M. Kinney, C. Steven, M. Rose, E. Karl, M. Weingarten, M. Karim Valji, B. Steven, M. Oglevie, C. Anne, M. Roberts, IVC Filter Tilt and Asymmetry: Comparison of the over-the-Wire Stainless-Steel and Titanium. J. Vasc. Interv. Radiol. **8**, 1029–1037 (1997)
73. Z. Jing, H. Mao, W. Dai, *Endovascular Surgery and Devices* (Springer Singapore, Singapore, 2018)

Chapter 5
3D Topological Scanning and Multi-material Additive Manufacturing for Facial Prosthesis Development

Mazher I. Mohammed, Joseph Tatineni, Brenton Cadd, Greg Peart, and Ian Gibson

5.1 Introduction

In a human population, facial defects can arise as a result of congenital deformities, disease infiltration, and trauma. Given the prominence of the face and how it influences human interactions, such disfigurements can have a profoundly negative impact on quality of life, often requiring repeated surgical interventions that aim on improving aesthetic appeal. With respect to rehabilitation, there are primarily two treatment options comprising either surgical intervention or the use of a prosthesis. The decision-making process over which option is the most suitable is not so clearly defined and dependent on a number of factors, ranging across the size/severity of the condition, age and aetiology as well as the patient's own personal preference [1, 2]. When considering prosthesis-based rehabilitation, there are several immediate advantages when compared with surgical intervention such as the immediate aesthetic improvement, its simplicity over surgery and consequently the reduced risk to the patient, the ability to explore numerous design iterations without impacting the patient and its comparatively low cost. More recently, there has been a surge of interest in tissue engineering approaches to replace missing or compromised organs [3, 4]. Despite the obvious potential of this technology, there are still many issues to resolve before this is likely to become a mainstream approach. Therefore, prosthetic

M. I. Mohammed · J. Tatineni
School of Engineering, Deakin University, Geelong, VIC, Australia
e-mail: mazher.m@deakin.edu.au; dawn.joseph@deakin.edu.au

B. Cadd · G. Peart
Facial Prosthetics, The Royal Melbourne Hospital, Parkville, VIC, Australia
e-mail: brenton.cadd@mh.org.au; greg.peart@mh.org.au

I. Gibson (✉)
Department of Design, Production & Management, University of Twente, Enschede, The Netherlands
e-mail: i.gibson@utwente.nl

P. J. Bártolo, B. Bidanda (eds.), *Bio-Materials and Prototyping Applications in Medicine*, https://doi.org/10.1007/978-3-030-35876-1_5

treatments provide a more robust, tried and tested approach which has a relatively quick and predictable turnaround time for part production, does not require extensive follow-up treatments and generally does not result in complications that are often associated with surgical interventions, such as tissue rejection.

Despite the advantages over surgery, traditional prosthesis production is still considered to be an unacceptably long and labour-intensive process, requiring the use of numerous invasive and subjective techniques throughout the fabrication process. A summary of the general fabrication stages can be found in Fig. 5.1. Typically the process begins with some form of casting using plaster to ascertain the topology of the defective area or of uncompromised anatomy that could be used as a template for the prosthesis [5, 6]. In some instances, plaster can be placed over the entirety of a patient's face, requiring breathing to be performed through a straw until the plaster sets. Additionally, due to the discomfort of this process, the patient may move during the casting, resulting in a subjective topological map. Following the formation of the plaster cast, a wax model is formed and manipulated to create the finished prosthesis model. At this stage, fixation point alignments may also be performed, which are embedded into the wax model. The alignment and finishing of the wax model are all performed manually, with the end result being wholly dependent on the art skill and artistic interpretation of the technician and is arguably very subjective in nature [7]. Once the wax model is finalised, it is further formed into a plaster negative, which in turn is used to generate the final silicone prosthesis. Any additional touches to the model, such as the addition of colours and hair, are again performed manually at this stage. It is clear to see that this whole process of prosthesis production is extremely arduous and labour intensive.

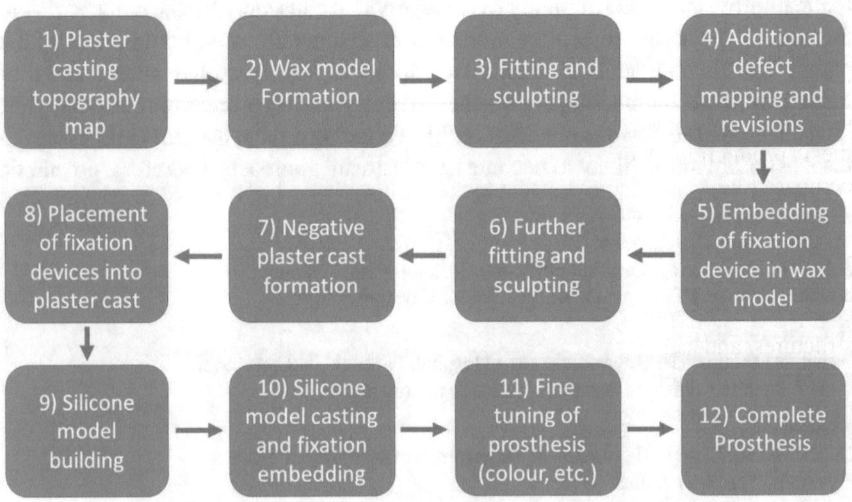

Fig. 5.1 Traditional process chain for external prosthesis fabrication

Traditional prosthesis development appears to be approaching a turning point, where several disruptive 3D technologies are likely to transform the previously mentioned process. Modern optical scanning technologies are now readily available and allow for the rapid, high-resolution reproduction of surface topologies to a precision of <100 μm whilst capturing useful data such as the texture of a patient's skin [8–10]. Additionally, such data is obtained non-invasively and can even be performed using laser-free methodologies that greatly improve the potential uptake in a clinical setting. Modern computer-aided design (CAD) software can readily manipulate this 3D scan data to create a complete model, with the added advantage of digitally recording the design iterations and storing the final model so that it may be used multiple times and in different contexts. Such digital data capture and storage, when applied to prosthesis production, compares very favourably to traditional hand-crafting techniques. Furthermore, modern additive manufacturing (AM, also referred to as 3D printing) technologies offer the ability to easily and rapidly reproduce the high-resolution digital model into a physical part [10–16]. Fabrication in this manner could be achieved using flexible and biocompatible materials, allowing for augmentation of existing practises and potentially could be applied to direct prosthesis fabrication.

In this study, we have investigated the use of optical scanning, reverse engineering, CAD and additive manufacturing towards the direct production of various facial prostheses (ear and nose). Previous studies have focused on the use of such technologies for the production of castings to augment traditional techniques or for parts made in rigid plastics. By contrast, the novelty in this work is the use of high-resolution AM with graded flexible material capability for the direct production of a prosthesis which not only results in a higher-quality surface finish, compared with the previous studies, but also mimics the tactile feel and pigmentation of human soft tissue. We also investigate the production of advanced prosthetic models, comprising multi-material designs, which more closely reproduce human tissue through mimicry of both the skin and cartilage. Our technique offers several advantages over traditional approaches such as the use of optical scanning for topological mapping. This approach is non-invasive, can acquire data within minutes and realises anatomically precise, high-resolution data sets that are ideal for prosthesis production. A digital CAD approach for prosthesis design is superior when compared with traditional casting and handcrafting because design iterations can be easily digitally stored, do not require any fabrication/material consumption and allow for operations such as mirroring of the original data set. Finally, the use of high-precision additive manufacturing allows for rapid digital part realisation and can produce a sophisticated prosthesis that comprises complex multi-material structures and can further allow for precise reproduction should duplicates be required. Ultimately, we believe the techniques presented in this work can realise a patient specific, low-cost and high-resolution approach to streamlining prosthesis optimisation and production.

5.2 Experimental

In this study, we have investigated the use of several 3D digitising, rendering and printing technologies to directly create prosthesis replica from a person's anatomy. The complete process chain can be seen in Fig. 5.2, where surface topology maps are made using an optical scanner, designs are post-processed using reverse engineering software followed by CAD and the final model is realised using high-resolution, multi-material AM.

5.2.1 Topology Mapping and Model Construction

In this study, a laser-free, optical scanning system (Spider, Artec, Luxembourg) was employed to obtain the surface topology of the nose and ear of a volunteer subject. Laser-free scanning technology alleviated any health concerns resulting from laser exposure to the eyes. The scanner used had an image scan resolution of approximately 50–100 μm, which is more than adequate to resolve all the major and minor details of the anatomical part rendered. The scanner operated alongside a proprietary software (Artec Studio 10, Artec, Luxembourg), which allows for the real-time visualisation of the scan data during acquisition. It was found that several translations of the scanner were required to obtain the full surface map, comprising movements at approximately 10 cm/s in a lateral and vertical arcing motion. The scanner allowed for the regions of interest to be captured and rendered rapidly within approximately 3–5 minutes.

The scanner software processes the input data as a point cloud, which is then converted into a full contour map by simply joining these points together using adjacent vectors. Rudimentary operations can be performed within the Artec software to condition the data, remove spurious noise, crop unrequired data regions, fill

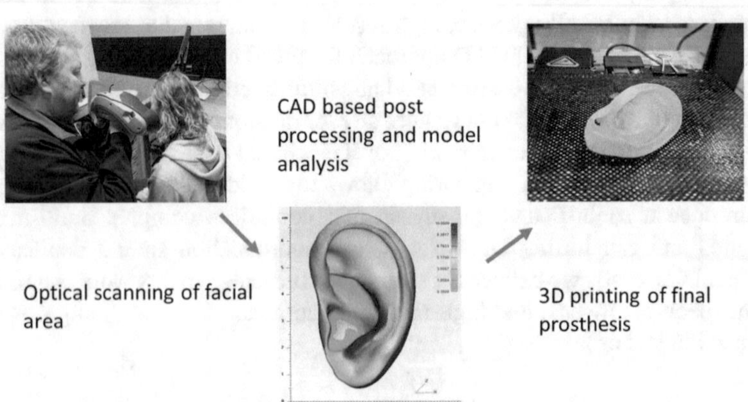

Fig. 5.2 Diagram illustrating the proposed process chain for prosthesis production

holes/gaps in the model and smooth contours. The resulting model appears in a form that is a completely enclosed model exported in the industry standard STL file format for AM. However, further post-processing may be required to the hollow part for the purposes of lightweighting, for example, and realise features such as nasal cavities and ear canals. Such features may correlate to regions where the scanner cannot cover due to overshadowing of other features, for example. In most cases, it would not be appropriate to push or pull on the patient to enable the scanning process as the intention would be to capture the surfaces in their natural state. This may be particularly problematic around the back of the ear, for example, although it should normally be fine to pin back the hair of the subject.

5.2.2 Model Construction and Post-processing

Data from the scanner was further conditioned using more sophisticated CAD software to process the surface topological data and to construct more advanced models for composite model printing. In this study, all additional post-processing was performed using the 3-Matic software package (Materialise, Belgium), which can allow for direct STL data manipulation, error checking and geometric measurements such as part thickness analysis. Data conditioning in our approach comprised smoothing of apparently rough surfaces, removing features that could lead to failure of the model when printing and performing procedures such as part hollowing. The need to smoothen rough surfaces is a particularly subjective decision. It may be that the surfaces are slightly corrupted by reflective artefacts. It may also be that the scanned surfaces are in fact rough due to scar tissue or other blemishes. The degree and regions for smoothing would therefore be a clinical decision.

Some of the scanning techniques had previously been developed in conjunction with the facial prosthetics group at the Royal Melbourne Hospital [17]. This study used AM to create a pattern based on scan data which was then manually manipulated to create a final silicone part for the patient. The major benefits of this approach lay in the speed and convenience of the scanning plus the accuracy of the pattern. In this further study, we aimed to realise an advanced prosthesis using AM that closely mimics human physiology in terms of tactile feel and pigmentation. As a demonstrator, we took a constructed model of the ear using the scanning approach earlier described. We then manipulated it to build separate models of the cartilage and a composite of the softer tissues. The regions representing the cartilage model were constructed using purely a design-based approach, selecting the regions within the data set using tools in 3-Matic. To obtain anatomical accuracy, the model was constructed by cross-referencing anatomical drawings of the ear cartilage and correlating that with qualitative measurements from direct feel of the test subject's actual ear. Approximations were then used to determine the layer thickness of the skin relative to the cartilage in the model. Once this was done, the remaining data inside the skin was considered to be cartilage. These two model sets can then be referenced to different material (i.e. stiffness) properties.

5.2.3 Additive Manufacturing

Once rendering and post-processing had been completed, the final models were directly 3D printed to produce the final prosthesis. To obtain acceptable accuracy in digital reproduction, a high-resolution AM machine (printer) was used (Connex 3, Stratasys, USA) which can print models in up to three individual materials or blends thereof to an accuracy of 16 μm layer thickness and 30 μm in-plane resolution. This was considered to be more than adequate for the reproduction of all major and minor surface contours. It also realises a sufficiently high-quality surface finish for the prosthesis, similar to that obtained by traditional manual techniques.

Using AM machines such as the Connex 3, a model is loaded into the printer's software as an STL file and then allocated specific process parameters before being sliced into the individual layers for printing. Simultaneously, a water-soluble support material is automatically generated at this stage of the process and allocated to ensure the build integrity. The printer operates using PolyJet™ technology, whereby liquid photo-curable polymers are delivered in droplet form by a printhead and subsequently flattened and cured by a UV lamp. Once a build has been completed, a subsequent cleaning phase is required to remove the surrounding support material before the part is ready for use. The printing process can take on average 2–3 hours for a part the size of an ear or nose to be printed, depending the orientation and whether there are other parts being built at the same time.

The printer used in this study was capable of printing in several different materials, which have a variety of colours and mechanical properties. For instance, the Tango Plus™ material range is a flexible monochrome material, whilst Vero™ materials are rigid and coloured. For the final prosthesis parts, we examined a combination of both Tango plus and Vero materials such that we could obtain a final model with not only a compliant tactile feel but also with adjustable colours. It was our hope in this study to realise combinations that mimic the feel and pigmentation of a person's actual tissue. Further, the advanced multi-model approach allows for further control of the tactile feel of a given anatomical part.

5.3 Results

5.3.1 Scanning and Model Creation

Various scans were performed on a test subject in an attempt to reproduce the surface topology of their nose and left ear. In this instance, the subject was a healthy individual who suffered no significant injury or facial defects. This allowed for a relatively simple comparison of the printed prosthesis with the original anatomy. When performing scans, the Fast Fusion™ mode of the scanner was utilised, which allowed for rapid, real-time visualisation of the scans on the computer as they were being performed. Scans were carried out by translating the scanner through the various orientations previously described. Following completion of the data acquisition, several additional data conditioning phases were performed using Artec Studio to remove obvious unwanted regions and spurious noise and to reconstruct small gaps in the data. Following completion of the scans, the software allowed for several automated procedures to merge multiple scans together, crop the part of interest from the wider facial data and convert the scan from a surface to a fully enclosed mesh.

It was found that there were limitations to the scanning process, and areas of the skin which were shiny/reflective were difficult to render during scanning. This limitation could be overcome using a matt powder to dampen optical reflections. Additionally, regions which had low levels of light exposure or shadowing were equally difficult to render, such as the nasal cavities, behind the ear and the ear canal. As such features are critical to the final prosthesis, they had to be rendered independently using the 3-Matic software.

5.3.2 *Model Post-processing*

5.3.2.1 Single Model

Following initial rendering, the models were checked using the Artec software to remove errors such as inverted normal vectors, multiple shells, noisy shells, among others. This procedure ensures the best-quality digital data for the subsequent design phases. Qualitatively, the rendered models from the scanner looked reasonably close to the original anatomy and the final prosthesis models that were printed, and so in this case, only relatively simple post-processing was required. With respect to the nose data, the nasal cavities were completely enclosed, and so the first procedure was to remove excess digital material to open these areas and to provide access to the reverse side of the model. The necessity for this is that a recipient of such a prosthesis may still have use of their nasal ducts, and so open access here would allow for a potential patient to retain the ability to breathe, smell, etc. With respect to the ear data, by cross referencing the major contours against the original test subject, it was found that several of the contours formed by the cartilage had been lost during the post-processing using the Artec Software. These areas were subsequently manually reconstructed using the 'push/pull' and 'extrude' functions within 3-Matic. Figure 5.3a illustrates the final model of the ear prosthesis.

With respect to the ear, when a thickness analysis was performed, it was found that there were regions which were <350 μm thick. Whilst this may not be an issue to the aesthetics within the digital mode, should this be printed, these regions would be extremely fragile. Initial test prints with this thickness in the rubber, such as Tango Plus material, found that they ruptured during the support material removal phase. Several design iterations were examined, increasing the minimum thickness in steps of approximately 200 μm. It was found that with thicknesses

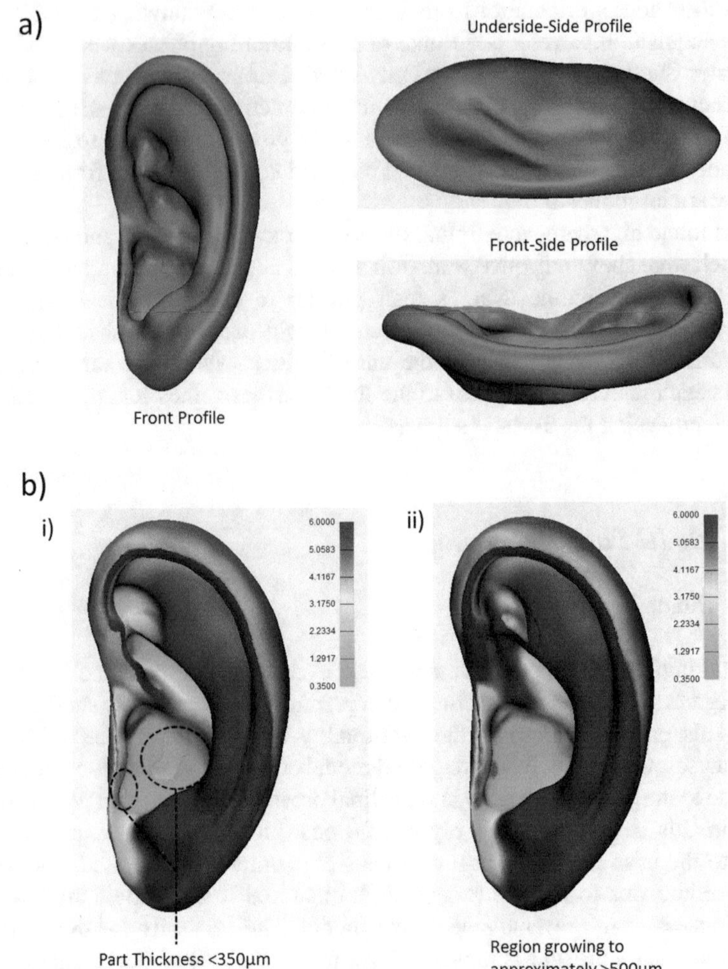

Fig. 5.3 (**a**) Various orientations of the ear prosthesis model. (**b**) Colour map thickness analysis of the ear model for (i) the raw data model input from the 3D scanner software and (ii) post-processed model to grow the thickness of the inner ear section to ≥1 mm

greater than 1 mm, the models could be cleaned without rupturing. Therefore, a thickness analysis of the part to be built is critical to the integrity of the final prosthesis using the Tango Plus material, and parts should be thoroughly inspected to ensure no regions exist with a thickness less than 1 mm. Figure 5.3b shows a model of the ear before and after the thickness analysis and subsequent model region growing.

5.3.2.2 Multicomponent Ear Modelling

The anatomy of the human ear comprises a single piece of skin-covered cartilage tissue in the ear which makes the design process relatively simple in comparison with more complex cartilage structures in organs such as the nose. Initially attempts were made to render the cartilage of the ear as a stand-alone model. To achieve this, the original ear model was duplicated and reduced in size by 96% to create a replica structure that was positioned approximately 1.5 mm into the original ear model. This new model was then reformed with reference to anatomical drawings of a generic human ear cartilage, the contours of the scanned ear model and by direct feel of the original human subject. Despite being a subjective metric for development of a design, the feel of the subject's ear provided valuable insight into the patient-specific anatomy of the ear, whilst the anatomical medical diagrams provided a framework for the general shape of the ear. The final model that was achieved can be seen in Fig. 5.4, which closely matched the medical diagrams.

5.3.3 Prosthesis Fabrication Using Additive Manufacturing

5.3.3.1 Single Model Printing

Following completion of the models, initial tests were performed to ascertain the printing precision, model integrity, mechanical conformity and pigmentations that could be achieved using a multi-material combinatory approach. The Connex printer is capable of printing with three materials simultaneously; however, as flexibility

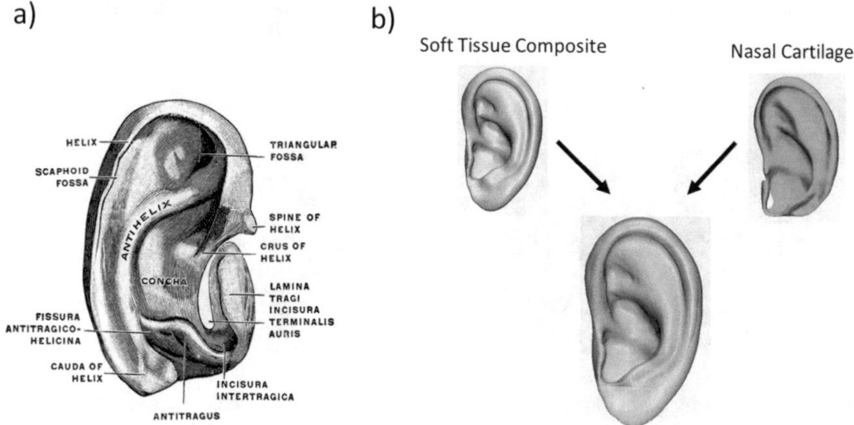

Fig. 5.4 (**a**) A classical medical diagram of the cartilage within the ear [18]. (Adapted from https://lookfordiagnosis.com/mesh_info.php?term=ear+cartilage&lang=1). (**b**) The multi-model design of the ear prosthesis comprising the soft tissue composite and the cartilage. For visualisation purposes both segments are also represented individually

was desirable through the use of Tango plus, only two additional materials (colours) could be used. Given the availability of colours by the manufacturers, it was decided that blends of magenta and yellow would provide the best options for skin pigmentation mimicry. A colour map was produced using the Connex machine of the different material combinations when using Tango plus and Vero materials, which can be seen in Fig. 5.5a. Another constraint that arose was the percentage blend of the rigid Vero with the flexible Tango plus material. Values of greater than 50–60% Vero material provided a tactile feel beyond the desired flexibility found in a typical prosthesis. The Tango plus material used in this study was a translucent variant, which on its own provided the softest tactile feel but was not a suitable colour for a prosthesis. It was found that a minimum of 10% Vero material was required to provide

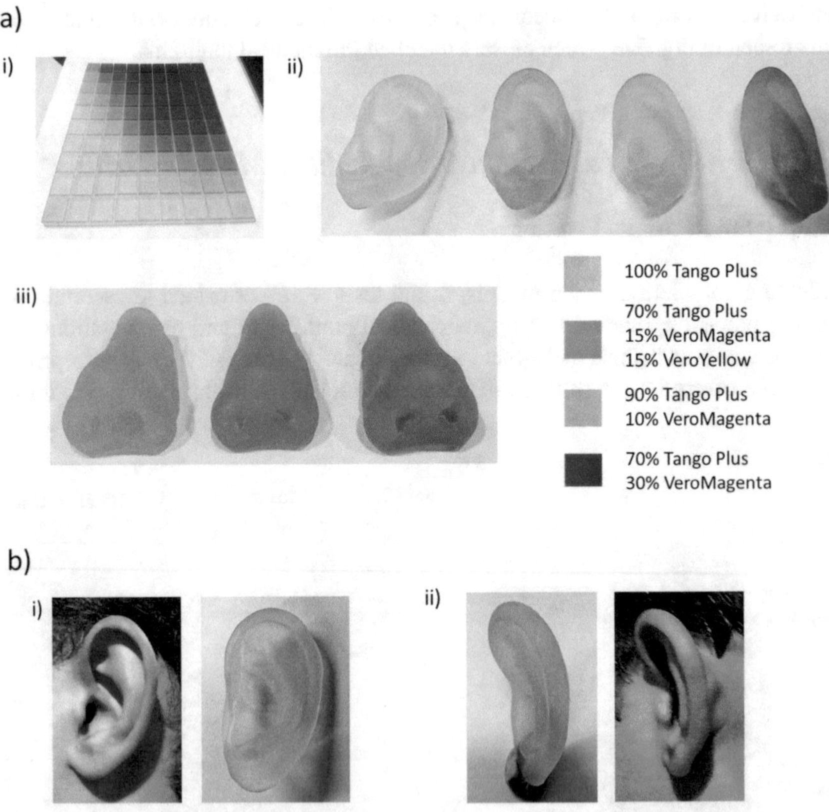

Fig. 5.5 (**a**) (i) A image of the material map which shows the various colour combinations possible with the 3D printer using two Vero colours and the flexible Tango plus material and images of various printed prosthesis of (ii) an ear and (**b**) a nose model. Material combinations are also highlighted in the legend. (**b**) Comparative images of the subject's ear and the printed model from (i) side and (ii) reverse orientations

any visually noticeable pigmentation into a given prosthesis. Various combinations of Tango and Vero material were explored to print the final prosthesis, and some of these can be seen in Fig. 5.5a.

The high resolution of the Connex printer and the use of dissolvable support material allows for the reproduction of both the ear and nose models with a smooth surface finish. The reproduced models were found to be a near identical match to the original anatomy, both in terms of major surface contours and dimensional deviations. Various nose and ear models can be seen in Fig. 5.5a (iii), showing clearly how the multi-material printing allows for a diverse array of prosthesis pigmentations from lighter to darker skin tones. Also shown in Fig. 5.5b are comparative images of the printed ear prosthesis next to the volunteer subject's original ear. It can be seen that the final prosthesis very closely matches the original contours of the ear, validating the efficacy of this technique. Upon closer examination of the ear models, it was found that the richness of the visible colour changed depending upon the thickness of the material. By comparison, the nose models failed to exhibit such features. This is perhaps due to the relative thickness ranges of these models. For the ear models, most thicknesses were between 1 and 6 mm, thus exhibiting varying degrees of translucency. By contrast, the minimum thickness of the nose was approximately 15 mm. At these thicknesses, all translucency effects are negligible. This is similar to human physiology, where the thickness of the soft tissues overlaying the cartilage of the ear has a variety of different pigmentations owing to its relative thickness. We hope to examine such effects more closely in future work.

The printed models were qualitatively assessed for their tactile feel and mechanical compliance in comparison with the original human subject. Tests comprised the ability to flex the lobe area, a vertical compression test from the lobe to the upper portion of the ear and a lateral compression test pinching across the centre of the ear. It was found that the printed models all exhibited similar characteristic to the original anatomy and that upon relaxation of the compressive force, the ears would return to the neutral position, as can be seen in Fig. 5.6a. For this experiment, a range of models were created using a fixed ratio of flexible to rigid material throughout each one. It was noted that the printed material was still noticeably more rigid, despite its elasticity, than real human anatomy. This resulted in a greater force being applied to achieve the various modes of compression, highlighting limitation of current printable, flexible materials. It was also noted that as the percentage of the Vero material was increased in the model, the applied force became larger. At a percentage of 60% Vero, the elasticity of the model was compromised beyond an acceptable level and became noticeably rigid. This limits the wider colour combinations possible whilst still retaining acceptable levels of elasticity to mimic human mechanical properties. We hope in future work to quantify such forces both for real anatomy and printed prosthesis.

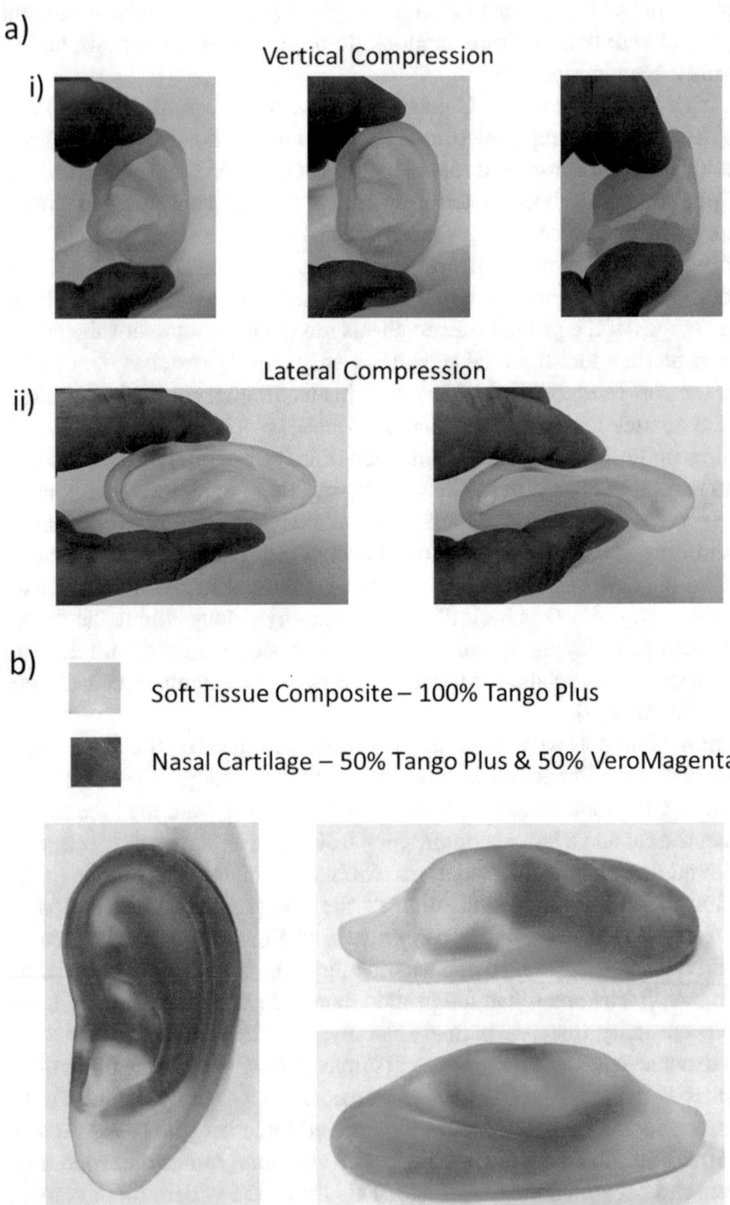

Fig. 5.6 (**a**) Qualitative compression of the model to demonstrate the realism in the tactile feel for both (i) vertical and (ii) lateral compression. (**b**) Multi-model printing of the ear, with independent material combinations for the cartilage and residual soft tissue composite

5.3.3.2 Multi-model Printing

As well as the previously mentioned single material models, we attempted to realise a range of models rendered using variable material combinations throughout. The purpose of this was to more closely mimic the softer and harder tissue regions of the ear. The Connex printer allows for the nesting of multiple models and for independent material allocation to each for printing. We therefore processed the advanced ear model such that the soft tissue composite was printed with 100% Tango plus and the encapsulated cartilage model was printed with 50% Tango plus and 50% VeroMagenta. These material combinations were used primarily for visualisation purposes such that each element could be visually differentiated, and the final printed model can be seen in Fig. 5.6b. The Connex printer was found to provide excellent multi-material printability, with the two materials seamlessly blending into a single structure without any impact on the final surface finish of the model. Therefore, the PolyJet printing process of the Connex is considered ideal for rendering of not only blended materials but also multiple material models for direct prosthesis printing.

Once again, the mechanical properties of the model were assessed by qualitative compression tests. On this occasion, it was found that there was much greater rigidity to the ear model, as expected; however, the ability to be compressed vertically and laterally was not compromised, and the model would also return to the neutral position upon relaxation of the compressive force. It was noted that the movement of the ear lobe was identical to the single material/model prints. Outcomes of the multi-model printing illustrate that the mechanical properties of the printed models can be further modified to reach increasing levels of complexity, as found in actual human anatomy. We hope in the future to more quantitatively assess the mechanical properties and to explore the ability to more seamlessly blend the colour combinations into a more realistic final model.

5.4 Conclusions

This study has demonstrated the potential of the 3D design and multi-model/material printing, augmented with the use of optical surface scanning, to produce realistic prosthetic models of both the ear and nose. The fabricated prostheses were realised to a high degree of accuracy, and we believe the technique suitable to render additional prosthetic parts, such as for an orbital prosthesis. We realised advanced prosthesis models beyond the traditional single material variants using novel design techniques to render components such as cartilage alongside other soft tissues, such as the skin. Using the multi-material printing approach, we could tailor the skin pigmentation of the prosthesis to a variety of skin tones, whilst also mimicking the mechanical properties of the original anatomy. The mechanical properties can be further tailored using the multi-model, multi-material approach used in this study. Currently, there are limitations in the complexity of the skin tones that can be mimicked without compromising the mechanical properties, due to the percentage material combinations available with the machine used. Additionally, the overall printed prosthesis tactile feel is more rigid than actual anatomy. These limitations are primarily due to the materials used in the 3D printer, but we believe as the technology matures over the coming years, these limitations will be resolved. For example, this is something that the recently introduced Stratasys J750 machine should be able to overcome, since the extra printhead permits the use of all colours with the addition of material hardness variations. Ultimately, the findings in the work validate our approach for direct prosthesis production which overcome limitations relating to the subjective nature of current prosthesis fabrication and allow for production within a single day. This compares favourably to traditional techniques where typically a prosthesis is fabricated over several weeks/months. Ultimately, this technique holds considerable potential for implementation within a clinical setting, streamlining the overall process for prosthesis production and seeing applications in other niche areas such as soft robotics or anatomical modelling.

This chapter is an adaptation of a paper presented at the 27th Solid Freeform Fabrication Symposium in Austin, Texas, USA, 2017 [19].

Acknowledgements We would like to thank the Royal Melbourne hospital for their input on the clinical aspects of this project. We would also like to thank the School of Engineering at Deakin University who provided funds and resources for this pilot project and their technical staff who assisted in the additive manufacturing of the models.

References

1. J.D. Kretlow, A.J. McKnight, S.A. Izaddoost, Facial soft tissue trauma. Semin. Plast. Surg. **24**(4), 348–356 (2010)
2. K. Ranganath, H.R. Hemanth Kumar, The correction of post-traumatic pan facial residual deformity. J. Maxillofac. Oral Surg. **10**(1), 20–24 (2011)
3. M.S. Mannoor et al., 3D printed bionic ears. Nano Lett. **13**(6), 2634–2639 (2013)
4. L. Jung-Seob et al., 3D printing of composite tissue with complex shape applied to ear regeneration. Biofabrication **6**(2), 024103 (2014)
5. S.S. Mantri, R.U. Thombre, D. Pallavi, Prosthodontic rehabilitation of a patient with bilateral auricular deformity. J. Adv. Prosthodont. **3**(2), 101–105 (2011)
6. A.N. Ozturk, A. Usumez, Z. Tosun, Implant-retained auricular prosthesis: a case report. Eur. J. Dent. **4**(1), 71–74 (2010)
7. R.M. Jani, N.G. Schaaf, An evaluation of facial prostheses. J. Prosthet. Dent. **39**(5), 546–550 (1978)
8. L. Ciocca et al., CAD/CAM ear model and virtual construction of the mold. J. Prosthet. Dent. **98**(5), 339–343 (2007)
9. G. Sansoni et al., 3D imaging acquisition, modeling and prototyping for facial defects reconstruction. SPIE Proc. **7239**, 1–8 (2009)
10. D. Palousek, J. Rosicky, D. Koutny, Use of digital technologies for nasal prosthesis manufacturing. Prosthetics Orthot. Int. **38**, 171–175 (2014)
11. M. Al Mardini, C. Ercoli, G.N. Graser, A technique to produce a mirror-image wax pattern of an ear using rapid prototyping technology. J. Prosthet. Dent. **94**(2), 195–198 (2005)

12. E.J. Bos et al., Developing a parametric ear model for auricular reconstruction: a new step towards patient-specific implants. J. Cranio-Maxillofac. Surg. **43**(3), 390–395 (2015)
13. Y. He, G.H. Xue, J.Z. Fu, Fabrication of low cost soft tissue prostheses with the desktop 3D printer. Sci. Rep. **4**, 6973 (2014)
14. I. Kuru et al., A 3D-printed functioning anatomical human middle ear model. Hear. Res. **340**, 204–213 (2016)
15. P. Liacouras et al., Designing and manufacturing an auricular prosthesis using computed tomography, 3-dimensional photographic imaging, and additive manufacturing: a clinical report. J. Prosthet. Dent. **105**(2), 78–82 (2011)
16. K. Subburaj et al., Rapid development of auricular prosthesis using CAD and rapid prototyping technologies. Int. J. Oral Maxillofac. Surg. **36**(10), 938–943 (2007)
17. https://www.youtube.com/watch?v=I5YthHUA9T0&t=61s
18. https://lookfordiagnosis.com/mesh_info.php?term=ear+cartilage&lang=1
19. M.I. Mohammed, J. Tatineni, B. Cadd, P. Peart, I. Gibson, *Applications of 3D topography scanning and multi-material additive manufacturing for facial prosthesis development and production.* Proceedings of the 27th annual international solid freeform fabrication symposium, 2016, p. 1695–1707

Chapter 6
Medical Applications of Additive Manufacturing

Zhaohui Geng and Bopaya Bidanda

6.1 Introduction

Additive manufacturing (AM), also known as 3D printing or rapid prototyping, is an advanced and transformative manufacturing technology that can directly create 3-D objects from digital models. The name *Additive Manufacturing* comes from the processes that create a solid object through adding successive thin layers of material. Due to this additive nature, AM enables fabrication of highly customizable and geometrically complex products using a variety of materials with a relatively low cost and high versatility. After it was first introduced by Charles Hull in 1980s [9], AM processes quickly gained popularity in the defense, automobile and aviation sectors to create part prototypes for design improvement and validation. With the recent improvement of accuracy and expansion of printable materials, AM could also make finished products for much wider industrial applications.

One of the most exciting and life changing applications of AM is in the field of medicine. The popularity of medical products, such as surgical guide, robots, prosthetics, etc., have been long suffering from their steep price tag and inaccessibility. The ability of directly fabricate a physical model from a digital model allows economic, small batch manufacturing of these medical products and also makes it easier for physicians or medical researchers, since they do not need to worry about the complexity of manufacturing process design and planning. This, in turns shortens throughput times for fabrication leading to better patient outcomes.

There are many ways to classify medical applications of AM, either by the type of AM process (stereolithography (SLA), selective laser sintering (SLS), inkjet printing, fused deposition modeling (FDM), extrusion, etc.) or based on body parts, etc. In this chapter, we adopt a classification based on the application type, since it

Z. Geng (✉) · B. Bidanda
Department of Industrial Engineering, University of Pittsburgh, Pittsburgh, PA, USA
e-mail: zhg17@pitt.edu; bidanda@pitt.edu

P. J. Bártolo, B. Bidanda (eds.), *Bio-Materials and Prototyping Applications in Medicine*, https://doi.org/10.1007/978-3-030-35876-1_6

will not only provide an overview of the current-state-of-the-art medical applications of AM, but will also be helpful to the reader to identify current gaps existing in the medical applications. According to Tuomi [21], the various application types include:

- Medical models for pre- and post-operative planning, education, and training. Through medical imaging and reverse engineering (RE), anatomical objects can be recreated with tactile quality for better and more accurate surgical planning and simulation.
- Medical aids, orthoses, splints, and prostheses. AM could provide personalized device to enhance healing and help damage repair as accurate as possible.
- Tools, instruments, and parts for medical devices. AM can help to create patient-specific tools to enable and improve efficiency of the medical procedures, and, also, patient-specific drug production.
- Inert implants: AM, together with medical imaging and RE, could provide customized implants for tissue replacement with an acceptable esthetic outcome, and biocompatibility material with good mechanical properties.
- Bioprinting: AM could produce freeform culture media through tissue engineering, that the biologically active tissue can be used for various purposes, such as drug screening and delivery, etc.

In this chapter, we first provide an overview of the entire AM process widely adopted by medical community and brief detail each step. Then, current-state-of-the-art medical applications of AM will be presented, while answering the questions such as AM's role in medical treatment, a comparison of traditional manufacturing processes versus AM. Next, the limitations of AM and the considerations for the medical applications of AM will be discussed. Finally, future trends of medical applications of AM will be presented.

6.2 AM Process in Medicine

Broadly speaking, the AM process can be divided into four major steps: pre-printing, printing, post-printing, and clinical application (Fig. 6.1). Depending on the type of application(s), the major techniques implemented in each step in the process can be quite different.

6.2.1 Pre-Printing

Pre-printing is the first step of the AM process, and includes all preparation work for the actual printing. Even though there may be minor variations depending on application type, this step follows the pattern below:

- Diagnosis,
- Imaging and scanning,

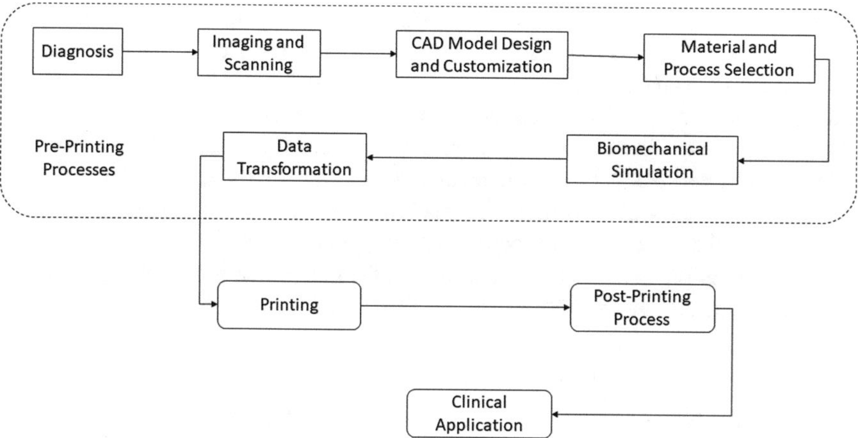

Fig. 6.1 Process flow of the medical additive manufacturing

- CAD model design and customization,
- Material and process selection,
- Biomechanical simulation, and
- Data transformation.

Diagnosis is the first step for any medical process and its purpose is to determine the objective and expected outcome of the AM application. Various medical tests can be used in this step, which is omitted in this chapter. Next is the step of imaging and scanning that is utilized to develop a digital model. Traditional medical applications either directly use 2-D computerized tomography (CT) scan or magnetic resonance imaging (MRI) images for diagnostics or use manual 3-D CAD modeling for manufacturing. However, due to the complex geometrical features and shapes of the biological objects, it is extremely hard, sometimes inaccessible, and always time consuming for medical diagnostics and to build patient-specific treatments. With the advancement of scanning and reverse engineering methodologies, 3-D CAD (re) construction of the biological objects can be much easier with a significantly higher degree of accuracy. Two of the most widely used imaging methodologies are CT and MRI due to their penetration capability to get internal layer images of both internal and surface structure without human body intrusion. 3-D scanners, such as coordinate measuring machines or laser scanners, are also used for medical applications, such as guidance or prosthetics production, but they can only scan the outer surface objects. After getting the data extracted by imaging and scanning, the CAD model is constructed by using surface fitting algorithms. Design and customization based on a CAD model can be implemented to get achieve medical objectives. For example, the physicians could add an internal structure for prosthetics to achieve better mechanical properties with lower weight, and dentists could use generated CAD models to provide critical surgical guidance for certain dental surgery [17] including sinus lifts.

The next step that includes material and process selection and bio-mechanical simulation is really iterative or can be considered as parallel, since based on the design, the physicians or medical researchers could list major properties that the final products must satisfy, such as mechanical properties, bio-compatibility, bio-degradability, etc. These properties dominate the materials and structures of design, which can be verified by bio-mechanical simulation. Also, the materials and properties of the design will determine the manufacturability of certain AM processes, which we will discuss in the following section. After the design is finalized, the 3-D CAD model is typically transformed to an STL file, which is traditionally, the de facto CAD file format for AM.

6.2.2 *Printing*

As part of the pre-printing process, the STL file is generated and passed to the appropriate AM machine. Even though the layer-by-layer printing of AM appears to be conceptually straightforward, various printing options and techniques are now available and each of them has different properties and material limitations due to their specific mechanism. In this chapter, we will not plan to investigate all available types of AM, rather, major AM technologies that are widely adopted by medical communities will be discussed.

6.2.2.1 Stereolithography (SLA)

SLA was developed in the mid-1980s by Charles (Chuck) Hull [9]. It makes use of radiation curable resins, or photopolymers, to create solid objects by additive successive 2-D layers that build on each other. The liquid materials undergo a sequence of photopolymerization process, which is a chemical curable reaction, to become a solid object (Fig. 6.2). Various types of radiation are available for SLA, such as gamma rays, X-rays, UV light, etc., while the materials are restricted to photopolymers, which can be expensive. SLAs generate accurate physical objects with smooth surface finishes, while due to the liquid nature of photopolymers, the support structure must however be carefully designed to hold the printed parts.

6.2.2.2 Selective Laser Sintering (SLS)

SLS, shown in Fig. 6.3, is also called, the powder bed fusion process and uses a CO_2 laser as the thermal source for inducing fusion between powder particles [2]. Instead of liquid resins as SLA, the thin layers of powder are spread across the build bed by a roller, which provide reasonable support to the printed parts. Thus, no support structures are needed to be considered in the design. Also, based on the fusion process, various materials can be used in this process, such as metal, plastic, and ceramic, etc. However, these fused small particles can cause rough surface finishes

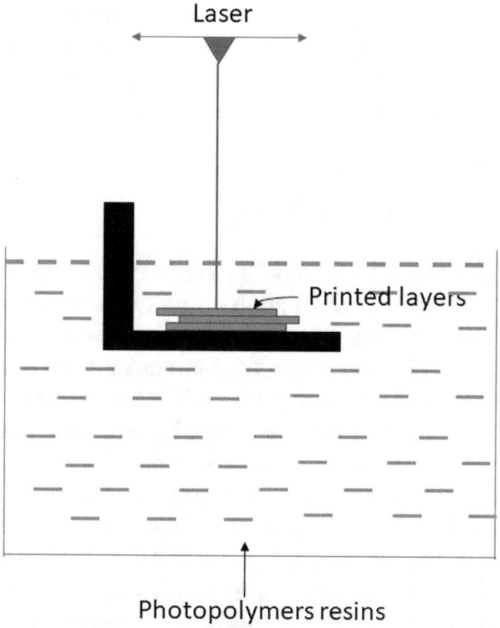

Fig. 6.2 The basic mechanism of the stereolithography (SLA) process

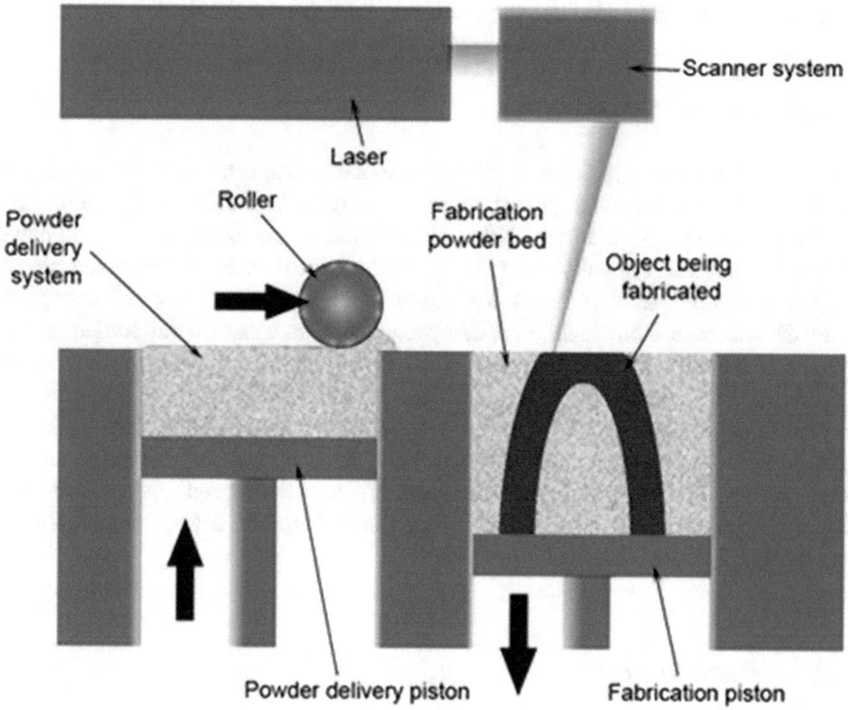

Fig. 6.3 The basic principle of the selective Laser Sintering (SLS) process. (Source: https://mechanicalstudents.com/additive-manufacturing-fused-deposition-modelling-fdm-selective-laser-sintering-sls/)

on the printed parts. Thus, the post-processing is often required for SLS process. Also, the powder bed is always preheated to improve the fusion process, which may cause the powder particles to become sticky and, further, drive the accuracy of the SLS process lower when compared to SLA and other AM techniques.

6.2.2.3 Inkjet Printing

Inkjet printing is one of the least expensive and most simple approaches of AM [18]. This process uses a printer head to create tiny ink droplets or continuous filament onto a powder bed, though ink droplets are more widely used in medical applications. After deposition, the inks are transformed and solidify to achieve desired properties of the finished products. The transformation is primarily due to the surrounding natural environment and the ink properties; however, postprocessing steps, such as heating, are also required. Since the inkjet printing is also based on power bed as the SLS, there is no need for a designed support structure. However, the introduction of ink materials eliminates the "fusion" process, and it the process, hurts accuracy.

Also, due to the simplicity of mechanism and flexibility, one exciting application of inkjet printing in medicine is to print organs. Instead of binding ink, the printer head can print living cells onto a biological scaffold to generate organs or cell cultures [15]. Furthermore, inkjet printing also provides the possibility of personalized medicine and accurate drug delivery, which will be discussed later.

6.2.2.4 Extrusion-Based System

The extrusion-based system is to force the material from the reservoir through a heated nozzle to generate a semi-solid state material filament, and fully solidify on the target surface [7]. Materials including acrylonitrile butadiene styrene and other thermoplastics are typically used for extrusion-based systems. One example of extrusion-based system is fused-deposition modeling (FDM). Overall speaking, extrusion-based system lacks fine detail and good mechanical properties as the other AM techniques and requires post-processing for good surface finishes. Also, unlike the SLS and inkjet printing, which are powder-bed based methodologies, extrusion-based system requires a support structure to prevent model collapse. However, the lower price of this system has gained it popularity among medical applications. Beyond printing biomaterial, an extrusion-based system can also directly print cells or multicellular system through the high-viscosity cell-laden hydrogels for various medical purposes.

6.2.3 Post-Printing

AM techniques, such as SLS, inkjet printing, extrusion-based system, often require post-processing steps. One major step is to smooth rough surfaces and remove excess material. For example, since the powder bed is preheated, the powder particles tend to be streaky in techniques, such as SLS and inkjet printing. Thus, instead of getting the printed parts, a powder cake is generated. An air blower or other equipment can be used to remove excess powder. Also, machining or surface polishing is required for certain applications to enhance biocompatibility or even for aesthetic reasons.

For specific applications, post-processing steps that increase strength and accuracy are required. For example, to print bone tissue, densification step is required in post-processing, since the sintered parts could cause shrinkage. More importantly, since the AM created parts are used for medical applications, sterilization is required to follow medical regulations and to maintain safe operating conditions.

6.2.4 Clinical Applications

Utilization in clinical applications is the ultimate goal of the AM process. After the AM created parts pass the regulatory requirements and approval, they are ready to be used in specific clinical applications, such as surgery, drug screening, etc. Clinical application can be very diverse. For example, in orthognathic surgery, a printed jawbone can be transplanted into the patient for jaw reconstruction. A 3D printed model enables surgeons to plan the liver surgery accurately and, even, locate tumors for removal. In maxillofacial surgery, the printed tissues can be implanted into patients by experienced physicians. When compared to traditional clinical applications, AM provides the capability to create more precise and patient-specific medical products and more effective surgical plans to reduce operation time and mitigate risks.

6.3 Medical Applications of AM

In this section, various medical applications are presented and classified based on their application. While it is obviously outside the scope of this chapter to cover every single medical application, we believe a representative sample of such applications that cover the large majority has been chosen.

6.3.1 Medical Models for Pre- and Post-Operative Planning, Education, and Training

A medical model provides the medical practitioner and researchers a better understanding of the biological objects of interests, since it is tactile, destructible, and more importantly, providing an accurate physical representation of the target. Traditionally, physicians work with 2-D images generated by X-ray, CT, or MRI to gain insight into pathologies. This process requires excellent visualization skills and rich experience from the physicians and/or surgeons. Also, since these 2-D images only represent cross-sectional images, they can be biased and time-consuming for a physician to interpret. Despite advancements in reverse engineering techniques that provide 3-D CAD models from CT and MRI, these models often lack tactile quality, making interpretation more time-consuming and has necessitate relatively high skill requirements. Precise replicas of complex anatomical model provide intuitive tactile views. With 3-D printed parts, surgeons are able to scrutinize every view and practice surgical procedures on parts, to reduce surgical planning time and improve treatment outcomes. Wurm et al. [24] use 3-D CT scans, 3-D rotational angiography, and SLA to create 13 solid anatomical bio-models of cerebral aneurysms with parent and surrounding vessels. Furthermore, Wurm et al. [25] further demonstrate that precise plastic replicas provide an accurate prediction of vascular anatomy for optimization of teaching and training surgical skill, surgical planning, and for patient counseling. Ahn et al. [3] propose orthopedic surgery planning based on AM and RE to improve efficiency and accuracy. Jiang et al. [10], use CT and AM generated models for cutting line planning of jawbone before cosmetic surgery in order to get a harmonious facial contour.

One of the major applications is surgical planning for liver transplantation. Due to the limited number of cadaveric livers or the complexity and high risk of liver transplantation from healthy donors, the success of liver transplantation surgery becomes very important. Traditional 3-D imaging may not meet the need for accurate planning and risk mitigation, since the 3-D image has to be projected onto 2-D screen, which requires excellent visualization and imagination capability. Thus, the perception of 3-D data can be quite difficult. With the help of AM, the surgeons could get a physical 3-D model from 3-D imaging, which can be viewed from various angles and provides better surgical plans to mitigate risks. Moreover, the 3-D model can also help surgeons locate tumors to plan surgery for tumor removal.

On the other hand, AM can be used in the development of medical student clinical competencies. Bringing anatomical variations from the clinics into preclinical studies can clearly improve medical student understanding of internal and external human anatomy structure. However, such resources can be seen as a considerable challenge for medical schools for a multitude of reasons including financial, ethical, legal, etc. Medical models created through AM brings excitement and inexpensive solutions to such challenges. Even without clinical studies, students could study human anatomy, practice surgery, and observe tissue behaviors through accurate AM created human tissues. McMenamin et al. [12], use AM to reproduce prosected human cadavers and other anatomical specimens to facilitate dissection-based teaching in medical. Reference [1] present the success that AM is used in anatomy education in Australia's Macquarie University and Western Sydney University.

6.3.2 *Medical Aids, Orthoses, Splints, and Prostheses*

Surgical guides are one of the more popular medical applications of AM. Applications of AM-based surgical guides can be found in maxillofacial surgery, neurosurgery, dental surgery, spinal surgery, etc. Surgical guides are used when the surgeon wants the most accurate placement of the medical product/device. Traditional surgical guide manufacturing requires the patients to expose substantial amount of time under ionizing radiation, which can be harmful for human body, and long surgical time due to the inconvenience of the inaccurate surgical guides. Newer AM-based surgical guides use a 3-D scanner to avoid damaging essential parts. And increase patient-specific accuracy, which can further improve the treatment outcomes in a cost-effective manner. Also, a patient-specific or customized surgical guide will enhance surgical precision and reduce surgical time, which in turn will reduce the anesthesia time that is harmful for patients.

Due to the flexible nature of AM, it can be utilized to produce customized and patient-specific prosthetics and orthotics. Instead of directly printing the CAD model generated from scanning and imaging, AM can be used to produce molds for making prosthetics. The major benefits of such prosthetics or orthoses or splints is to provide a geometrical design that will exactly match the original or preferred geometry and kinematics in order to improve patients' comfort, stability, and, furthermore, reduce the recovery and treatment time. More importantly, from patients' point of view, AM will help provide custom and precise prothesis at a lower price point. For example, the dental crowns can be created by AM based precisely on the current stage of the dental environment of the patients. The treatments could be more effective and the pain can also be heavily reduced. Also, the aesthetics and functionality of the prosthetics can be tailored based on patient-specific needs and wishes. Müller et al. [13] implement the craniofacial surgery using SLA to provide pleasing aesthetic outcomes for patients with craniofacial dysmorphism. Prosthetic sockets created by SLA are cost-effective and show better performance compared to machine- or hand-made ones [8]. Traditional manufacturing of a silicone ear can be costly and time consuming. Costal cartilage is even inaccessible to design and manufacture to a given, desired shape [16]. Turgut et al. [22], use AM and the mirror image of the contralateral ear to create an auricular prosthesis to the patients who lost one side of year due to tumors, congenital deformities. Their results show that AM could provide excellent forms with lower prices. Li et al. [11] propose to use CAD and AM in prosthetic chin augmentation for mild microgenia. Also, AM could produce inner structure for splints or prostheses that can both reduce the weight that provide the patients with more comfort and keep good mechanical properties.

6.3.3 Tools, Instruments, and Parts for Medical Devices

Traditional tools and instruments are produced in a 'one size fits all' fashion, that it can be applied to many patients and scenarios. Thus, the cost of the tools are relatively low, and surgical procedures are often planned according to the availability of certain tools. However, due to the individual difference of the patients, the entire surgical procedure can require a relatively long period of time and provide a less than satisfactory outcome. The flexibility of 3-D printing allows designers to make tools and instruments according to patient-specific dimension and shapes. Such tools improve the efficiency of the medical or surgical procedure. Due to the individual differences and highly patient-specific nature of the procedure, little research has been published in this field - however, this type of applications gains extensive popularity in medical industries. Gibson et al. [7], discuss two examples of in-the-ear hearing aids which are designed based on precise data from individual and basic generic design of product. The hearing aids are highly customizable and has components that suit a specific user. AM makes this type of application available for mass customization and easier for manufacturer and practitioners.

6.3.4 Inert Implants

Inert implants are widely used in cranial surgery, dentistry, and maxillofacial surgery. Most inert implants are related to the skeletal structure, which is the first and foremost mechanical structure of human body. The material at the bone-implant interface can be extremely important and subject to both chemical and mechanical limitations. The chemical composition of the material directly affects cellular reaction with the bone. Thus, biocompatible materials, such as titanium (Ti-6Al-4V), polyether ether ketone (PEEK), are widely adopted. In terms of mechanical properties, the unique structure of bone, which is not uniform and has an inner region with less density and an outer density with higher density. This makes the design of implants complicated and requires patient-specific and high accuracy of the implants. Thus, such implants based on traditional manufacturing are either time consuming or expensive, or even infeasible.

AM is utilized here to fabricate customized fixtures and implants with biocompatible materials. Due to its flexibility, AM is able to produce complex implants design that is suitable for the unique structure of bone and weight-compatible to the replaced bone. With AM, it is possible to create a precise implant for a specific patient, rather than standardized implants, which could provide the patients with better treatment outcomes. Custom or patient-specific implants created by AM have advantages of higher implant conformity, shorter surgery time, better functionality, faster healing, and lower mortality rates.

6.3.5 Bioprinting

Bioprinting is one of the most exciting and futuristic areas in medical AM. The ultimate goal of 3-D bioprinting technology is to directly create living and functional organs for various outcomes. Bioprinting applications can be subdivided into three major areas: pharmaceutics, cardiovascular applications, and cancer studies.

6.3.5.1 Pharmaceutics

Ideal drug dosing should be patient-specific, based on the patient's age, gender, weight, and body performance. Ideal drug dosing is also known as personalized medicine and is expensive using traditional drug manufacturing techniques because of the customized nature of the drug. AM techniques, such as inkjet printing [6], can be used to print a specific dose to achieve generate patient-specific drugs at a relatively low cost.

On the other hand, traditionally, drug screening has to use in vitro animal cell lines, while the conclusions made may not be applicable for normal human cells. Thus, it is attractive to create 3-D in vitro human tissue with multiple types of cells for drug screening and toxicology that mimic the in vivo behaviors. [4, 20]

6.3.5.2 Cardiovascular Applications

Bioprinting technology helps medical researchers create vascularized tissues and develop blood vessels and heart tissues based on 3-D models. This was simply not possible with traditional biomanufacturing techniques. However, Norotte et al. [14] have built biological vascular tubular grafts by employing bioprinting methodology that appears to be accurate, reliable, and scalable. More details on cardiovascular applications can be found in Shafiee and Atala [19], Chua and Yeong [5].

6.3.5.3 Cancer Studies

3-D bioprinting helps to create 3-D in vitro replica cultures that mimics the sophisticated cells-cell and cell-matrix interactions for cancer research. Thus, allows researchers to understand the crucial cancer mechanisms and develop novel clinical therapies. [23]

6.4 Limitations and Considerations

As we have seen in earlier sections, AM techniques have many advantages when compared to traditional manufacturing technologies. Major advantages can be summarized as follow:

- Reduction in surgery/treatment time
- Good accuracy
- Better medical outcomes
- Accurate and tactile anatomical representations
- Good representation of actual pathology
- Decrease of harmful exposure, such as ionizing radiation
- Enhanced collaborative efforts

However, there are still some limitations that can be improved to provide better AM performance in medical applications. These are briefly detailed below.

6.4.1 *Insufficient Accuracy*

There has been little work done on the overall accuracy of the AM process. Some researchers discuss the accuracy of specific AM techniques, however, as we defined in Sect. 6.2, the broader AM process regarding accuracy, starts with imaging and scanning and ends after post-processing. Considered as the first step, the accuracy of imaging technologies, such as CT, MRI, and 3-D scanners, cannot be fully modeled or understood using current-state-of-the-art methodologies, due to the complexity of the skeletal geometry. There is a research gap in judgmental methodology for complex 3-D modeling. Further, the geometrical file related steps that follow the imaging techniques, such as data transformation, STL file generation, printing, etc., can cause further deviations between the printed model and the true target object. Even though the deviations could be very small, it can cause significant trauma for patients in medical applications. Thus, further research in f tolerance design for AM is urgently needed for more effective AM applications.

6.4.2 *Increase in Cost Effectiveness*

Cost is one of the major deterministic factors in the usefulness and applicability of AM in clinical applications. Despite the drop in prices of commercial printers during past years, AM machines with high accuracy and resolution are still quite costly. Some medical institutions seek to outsource medical 3-D printing for specific parts, however, the patient's privacy issues and longer turnaround times could prevent the patients from getting timely treatment and increase costs for patients.

6.4.3 *Regulations and Safety Issues*

There are many critical questions regarding the safety of broad medical application of 3-D printing. The usage of AM in preoperative planning, education and training has not been proven to be effectiveness by researchers. Working in a data-rich environment that contains patients' privacy information, is somewhat risky due to concerns of the data privacy.

Furthermore, since we are still in the nascent stages of AM in medical applications, it is unclear how AM produced parts will affect the outcomes, since human body structures are extremely complex with many unknowns.

On the other hand, the regulatory limitations are one of the most challenging constraints to allowing AM to be used as a major technology for medical applications. As an emerging technology, there is no single category specified by US Food and Drug Administrations (FDA) that AM could fit into. Thus, the regulatory restrictions are urgently needed.

6.5 Conclusions

A broad range of medical applications using AM has been described. The entire AM process that are utilized in medical applications are discussed: pre-printing, printing, post-printing, and clinical application. Every step is equally important for medical applications, which requires highly skilled practitioners to participate to get desired treatment outputs. AM is a powerful and flexible tool for medical applications. It provides excitement to the medical community to provide patients with more suitable and reliable medical treatments and comfort. Further, several bottlenecks and restrictions of AM when applying to medical applications are also discussed.

Future research trends regarding AM in medical applications will likely be even broader. For example, in a pharmacy setting, one interesting possibility is the personalized drug as mentioned in Sect. 6.3.5.1. Instead of buying commercial standard drug, patients could ask their doctors to send personal information to the local pharmacy store and, through AM techniques, such as inkjet printing, the pharmacy could print personalized drug for the patients to get better and more efficient treatment.

Another fast-improving field of medical AM is bioprinting of complex organs. The ultimate goal of bioprinting is always to recreate a fully functional in vitro organs for various medical purposes. There is a huge research gap between current-state-of-the-art bioprinting and this ultimate goal and one might expect several decades before we reach this objective.

Current AM processes print 2-D material layers on either liquid resins or powder substrates, then the solid objects are moved to target treatment areas. Another ongoing research trend is in situ printing, that the printer directly print material onto the treatment area during surgery. This allows the user to create better mechanical and chemical properties to the treatment outcomes.

References

1. Y. Abou Hashem, M. Dayal, S. Savanah, G. Štrkalj, The application of 3D printing in anatomy education. Med. Educ. Online **20**(1), 29847 (2015)
2. M. Agarwala, D. Bourell, J. Beaman, H. Marcus, J. Barlow, Direct selective laser sintering of metals. Rapid Prototyp. J. **1**(1), 26–36 (1995)
3. D.G. Ahn, J.Y. Lee, D.Y. Yang, Rapid prototyping and reverse engineering application for orthopedic surgery planning. J. Mech. Sci. Technol. **20**(1), 19 (2006)
4. R. Chang, K. Emami, H. Wu, W. Sun, Biofabrication of a three-dimensional liver micro-organ as an in vitro drug metabolism model. Biofabrication **2**(4), 045004 (2010)
5. C.K. Chua, W.Y. Yeong, *Bioprinting: Principles and Applications*, vol 1 (World Scientific Publishing Co Inc, Singapore, 2014)
6. N. Genina, D. Fors, H. Vakili, P. Ihalainen, L. Pohjala, H. Ehlers, I. Kassamakov, E. Haeggström, P. Vuorela, J. Peltonen, N. Sandler, Tailoring controlled-release oral dosage forms by combining inkjet and flexographic printing techniques. Eur. J. Pharm. Sci. **47**(3), 615–623 (2012)
7. I. Gibson, D.W. Rosen, B. Stucker, *Additive Manufacturing Technologies*, vol 17 (Springer, New York, 2014)

8. N. Herbert, D. Simpson, W.D. Spence, W. Ion, A preliminary investigation into the development of 3-D printing of prosthetic sockets. J. Rehabil. Res. Dev. **42**(2), 141 (2005)
9. C.W. Hull, U.S. Patent 4,575,330, 1986. Washington, DC: U.S. Patent and Trademark Office
10. N. Jiang, Y. Hsu, A. Khadka, J. Hu, D. Wang, Q. Wang, J. Li, Total or partial inferior border ostectomy for mandibular contouring: Indications and outcomes. J. Cranio-Maxillofac. Surg. **40**(8), e277–e284 (2012)
11. M. Li, X. Lin, Y. Xu, The application of rapid prototyping technique in chin augmentation. Aesthet. Plast. Surg. **34**(2), 172–178 (2010)
12. P.G. McMenamin, M.R. Quayle, C.R. McHenry, J.W. Adams, The production of anatomical teaching resources using three-dimensional (3D) printing technology. Anat. Sci. Educ. **7**(6), 479–486 (2014)
13. A. Müller, K.G. Krishnan, E. Uhl, G. Mast, The application of rapid prototyping techniques in cranial reconstruction and preoperative planning in neurosurgery. J Craniofac Surg **14**(6), 899–914 (2003)
14. C. Norotte, F.S. Marga, L.E. Niklason, G. Forgacs, Scaffold-free vascular tissue engineering using bioprinting. Biomaterials **30**(30), 5910–5917 (2009)
15. F.J. O'brien, Biomaterials & scaffolds for tissue engineering. Mater. Today **14**(3), 88–95 (2011)
16. A.J. Reiffel, C. Kafka, K.A. Hernandez, S. Popa, J.L. Perez, S. Zhou, S. Pramanik, B.N. Brown, W.S. Ryu, L.J. Bonassar, J.A. Spector, High-fidelity tissue engineering of patient-specific auricles for reconstruction of pediatric microtia and other auricular deformities. PLoS One **8**(2), e56506 (2013)
17. M. Salmi, *Medical applications of additive manufacturing in surgery and dental care*, Ph.D. dissertation, Aalto University (2013)
18. H. Sirringhaus, T. Kawase, R.H. Friend, T. Shimoda, M. Inbasekaran, W. Wu, E.P. Woo, High-resolution inkjet printing of all-polymer transistor circuits. Science **290**(5499), 2123–2126 (2000)
19. A. Shafiee, A. Atala, Printing technologies for medical applications. Trends Mol. Med. **22**(3), 254–265 (2016)
20. J.E. Snyder, Q. Hamid, C. Wang, R. Chang, K. Emami, H. Wu, W. Sun, Bioprinting cell-laden matrigel for radioprotection study of liver by pro-drug conversion in a dual-tissue microfluidic chip. Biofabrication **3**(3), 034112 (2011)
21. J. Tuomi, K.S. Paloheimo, J. Vehviläinen, R. Björkstrand, M. Salmi, E. Huotilainen, R. Kontio, S. Rouse, I. Gibson, A.A. Mäkitie, A novel classification and online platform for planning and documentation of medical applications of additive manufacturing. Surg. Innov. **21**(6), 553–559 (2014)
22. G. Turgut, B. Sacak, K. Kran, L. Bas, Use of rapid prototyping in prosthetic auricular restoration. J Craniofac Surg **20**(2), 321–325 (2009)
23. C. Wang, Z. Tang, Y. Zhao, R. Yao, L. Li, W. Sun, Three-dimensional in vitro cancer models: A short review. Biofabrication **6**(2), 022001 (2014)
24. G. Wurm, B. Tomancok, P. Pogady, K. Holl, J. Trenkler, Cerebrovascular stereolithographic biomodeling for aneurysm surgery. J. Neurosurg. **100**(1), 139–145 (2004)
25. G. Wurm, M. Lehner, B. Tomancok, R. Kleiser, K. Nussbaumer, Cerebrovascular biomodeling for aneurysm surgery: Simulation-based training by means of rapid prototyping technologies. Surg. Innov. **18**(3), 294–306 (2011)

Chapter 7
Engineering Natural-Based Photocrosslinkable Hydrogels for Cartilage Applications

Hussein Mishbak, Cian Vyas, Glen Cooper, Chris Peach, Rúben F. Pereira, and Paulo Jorge Bártolo

7.1 Introduction

The desire to regenerate and replace damaged or dysfunctional human tissues to improve quality of life is as old as human history. This aspiration has been expressed in numerous cultures throughout history such as the myth of Prometheus with his eternally regenerating liver and Mary Shelley's 'Frankenstein' in which the scientist created new life from the rejuvenation of dead tissue. This human fascination to regenerate the body has in recent decades developed into the rapidly expanding scientific field of tissue engineering (TE). TE holds the promise to offer a paradigm

H. Mishbak
Biomedical Engineering Department, College of Engineering, University of Thi-Qar, Thi-Qar, Iraq

Department of Mechanical, Aerospace and Civil Engineering, University of Manchester, Manchester, UK
e-mail: Hussein.al-hasani@postgard.manchester.ac.uk

C. Vyas · G. Cooper · P. J. Bártolo (✉)
Department of Mechanical, Aerospace and Civil Engineering, University of Manchester, Manchester, UK
e-mail: cian.vyas@postgrad.manchester.ac.uk; glen.cooper@manchester.ac.uk; paulojorge.dasilvabartolo@manchester.ac.uk

C. Peach
Department of Mechanical, Aerospace and Civil Engineering, University of Manchester, Manchester, UK

Manchester University Foundation NHS Trust, Manchester, UK
e-mail: c.peach@doctors.org.uk

R. F. Pereira
i3S – Instituto de Investigação e Inovação em Saúde, Universidade do Porto, Porto, Portugal

INEB – Instituto Nacional de Engenharia Biomédica, Universidade do Porto, Porto, Portugal
e-mail: ruben.pereira@ineb.up.pt

P. J. Bártolo, B. Bidanda (eds.), *Bio-Materials and Prototyping Applications in Medicine*, https://doi.org/10.1007/978-3-030-35876-1_7

shift in clinical treatment and understanding of a host of disease states. This shift will enable a science fiction like prospect of infinite replacement human body parts which will have tremendous positive impact on human health. The concept of TE has developed into the currently accepted definition put forward by Langer and Vacanti in 1993 that states:

> "Tissue engineering is an interdisciplinary field that applies the principles of engineering and the life sciences toward the development of biological substitutes that restore, maintain, or improve tissue function." [1].

Due to an ageing population worldwide and increasing numbers of age-related and lifestyle linked diseases such as osteoarthritis, diabetes, and cardiovascular disease, the importance of TE is enormous. TE will enable the development of superior clinical treatments to the benefit of patients but also relieve the strain on healthcare systems. Furthermore, TE has applications beyond tissue regeneration and replacement such as the development of three-dimensional (3D) culture and tissue constructs that allow pharmaceutical testing, disease modelling, toxicology testing, and improved understanding of cell biology [2].

There are multiple strategies employed in TE to enable tissue regeneration and replacement. A scaffold based or top-down approach, typically utilises a scaffold which is a supporting construct that outlines the spatial requirement of the tissue, enables cell attachment, and provides environmental cues to promote a desired cellular response [3–5]. This approach can be typically split into either acellular or cellular techniques. Cellular approaches using a traditional in vitro TE strategy require cells to be isolated, expanded in culture, and then seeded onto a scaffold in vitro to enable maturation and remodelling of the construct before subsequent implantation into a patient. An alternative method is to seed the scaffold with cells isolated at the point of surgical implantation which avoids the in vitro maturation phase and requires a single surgical intervention. The acellular in vivo TE scaffold approach bypasses the need for cell seeding and maturation completely as a biofunctional scaffold is directly implanted into the patient. This approach relies on the recruitment of cells in vivo to initiate tissue repair and remodelling of the scaffold. An alternative approach is the scaffold-free strategy which is a bottom-up process utilising cell spheroids, cell sheets, and self-assembly to generate new tissue and typically does not require exogenous scaffolds or cell seeding [6]. These approaches utilise the ability of cells to fuse into larger constructs and in self-assembly necessitates the recapitulation of embryonic and developmental pathways of tissue and organ genesis [7–9]. The scaffold-free approach using cells, their own extracellular matrix (ECM) and signalling molecules aims to generate new tissues and organs with high level of biomimicry, thus requiring a deep understanding of the underlying biological and physical mechanisms of these pathways. Finally, a new hybrid approach aims to combine both scaffold and scaffold-free TE in a synergistic approach to benefit from the inherent advantages of both techniques [10].

Hydrogels are a specific class of materials highly relevant for both the fabrication of acellular scaffolds and cell-laden constructs. They are biocompatible and can be designed to be biodegradable with a structure that closely mimics the native in vivo ECM [11–13]. This is a result of hydrogels typically being composed of a three-dimensional (3D) polymer network which swells on contact with water [14]. This enables storage of large quantities of water which resembles soft biological tissues and allows for encapsulation of cells and bioactive molecules. Furthermore, hydrogels are typically biocompatible and allow for diffusion of oxygen, nutrients, and water-soluble compounds. Hydrogel properties can be tuned through chemical, physical, and biological modification which allows a variety of cell behaviours such as adhesion, proliferation, and differentiation to be optimised. Moreover, these 'smart' hydrogels can be tuned to respond to stimuli such as temperature, light, pH, and biomolecules which enables hydrogels to have a variety of properties such as self-healing, controlled degradation, and drug delivery [15]. A pioneering example of a cell-responsive hydrogel is Hubbell et al. demonstration that inclusion of specific peptides in a hydrogel could enhance matrix metalloproteinases (MMP) activity and thus control the rate of scaffold degradation [16]. The ability to fabricate hydrogels with specific properties has enabled precise control of cell behaviour which is a benefit in creating scaffolds that will allow tissue engineering to be successfully translated.

Photocrosslinkable or photocurable hydrogels are particularly relevant for TE applications due to the rapid gelation, control over spatiotemporal formation, and the ability to tune polymer network properties such as crosslinking density and matrix stiffness [17]. This is often achieved by controlling operating parameters such as light intensity, exposure time, and illumination area. When used in combination with bioprinting techniques, photopolymerisation enables the fast and automated crosslinking of complex scaffold architectures during and post-printing. However, photocrosslinking conditions need to be optimised to overcome concerns regarding cell exposure to UV radiation and potential cytotoxicity from photoinitiators, by-products, and unreacted reagents.

The aim of this chapter is to outline the recent advances in the design of naturally derived photocrosslinkable hydrogels for cartilage tissue engineering. A brief description of the structure and composition of cartilage tissue will be firstly provided, followed by an overview of hydrogel properties, synthesis routes, and strategies for hydrogel biofunctionalisation within cartilage tissue engineering (Fig. 7.1).

7.2 Articular Cartilage

Articular cartilage is a load-bearing and non-vascularised tissue that covers the ends of bone joints (Fig. 7.2a), acting as a low-friction bearing surface and a mechanical damper for the bones [19, 20]. It is a matrix-rich tissue with specialised cells (chondrocytes) that maintain the structural and functional integrity of the ECM [21–24]. However, cartilage lacks inherent self-repair ability and remains a significant clinical and economic challenge throughout the world following trauma and disease as the complex functional and zonal structure can be irreversibly lost. Cartilage defects are prone to develop into osteoarthritis (OA), which is a major disease associated

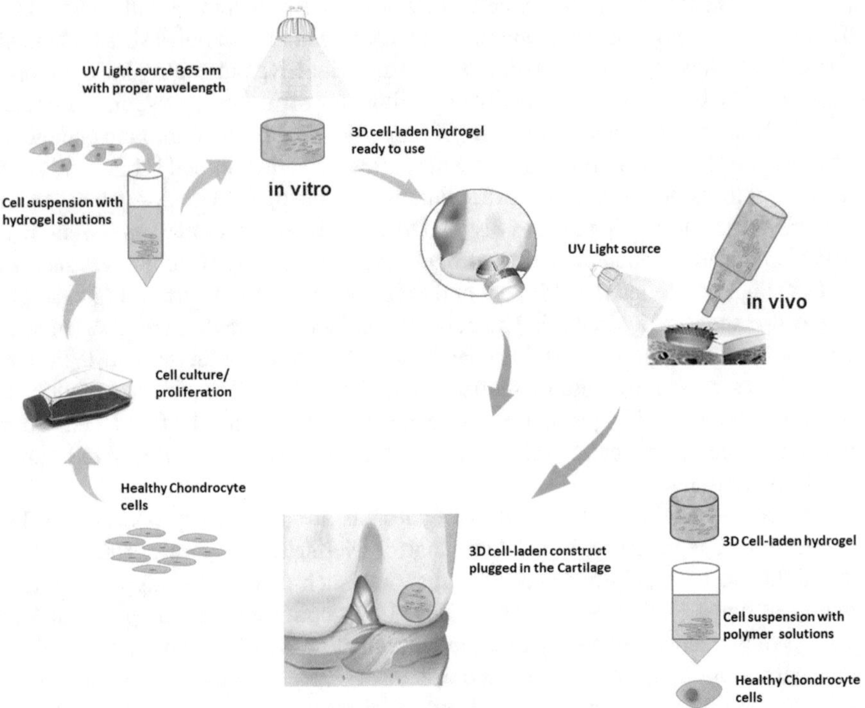

Fig. 7.1 Schematic diagram of photocrosslinkable 3D hydrogel constructs and approaches for use in cartilage tissue engineering

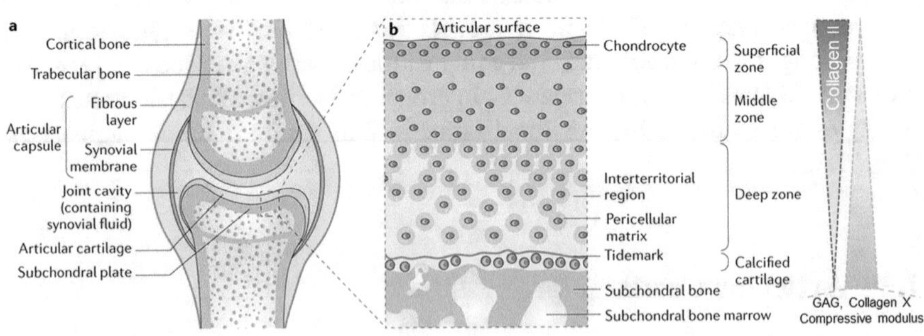

Fig. 7.2 (**a**) Articular cartilage covers the surfaces of bones in diarthrodial joints (e.g. knee and elbow). (**b**) Zonal structure and composition of articular cartilage [18]

with cartilage tissue and is characterised by a loss of healthy and functional cartilage tissue, pain, and debilitation thus is a significant clinical and economic challenge [25, 26].

7.2.1 Composition and Zonal Organisation

Cartilage is characterised by a distinct zonal structure comprising, in descending order, the articular or superficial surface, the middle (or transitional) zone, the deep zone, a tidemark (separates the non-calcified and calcified regions) [27, 28], and the calcified zone (Fig. 7.2b). These zones differ with respect to the molecular composition and organisation of the cartilage ECM, the shape, density, alignment of the resident chondrocytes and the mechanical properties of each zone. The subchondral bone acts as a shock absorber, the loading forces experienced by the joint are transmitted to this region, and the bone tissue contains blood vessels that supply nutrients, oxygen, and removes waste from the lower regions of the avascular cartilage tissue [29]. Surrounding each individual chondrocyte are three distinct ECM regions which depending on the distance from the cell are termed the pericellular, territorial, and interterritorial region. The highly organised and hierarchical properties of the tissue are responsible for its remarkable mechanical properties [28, 30–32].

Articular cartilage is typically composed of:

Chondrocytes responsible for production and maintenance of the cartilage ECM and essential for tissue formation and functionality. They represent about 5–10% of the total volume of cartilage [33, 34]. However, they have limited proliferation capability and low metabolic activity, which is partially responsible for the limited capacity of cartilage to recover from trauma or disease [35, 36].

Collagen the main component of cartilage with up to 60% of the dry weight being collagen with type II the most common type representing 90–95% of total collagen in the ECM which can form fibrils and fibres entwined with proteoglycan aggregates [37].

Proteoglycans molecules responsible for the resistance to compression of cartilage tissue [38]. Proteoglycans (PG), corresponding to around 35% of the cartilage dry weight, are made up of repeating disaccharide units, glycosaminoglycans (GAGs), with two main types; chondroitin sulphate and keratin sulphate [38]. Proteoglycans have a turnover rate of 3 months and weave between the collagen fibres creating a mesh responsible for retaining water and providing elasticity to the tissue [39].

Water represents 65–80% of the total cartilage mass [40]. This content reduces with aging and is linked to a reduction of strength and elasticity of the tissue [41].

Synovial fluid a highly viscous liquid secreted by synovial lining cells, they are responsible for secreting hyaluronic acid (the intrinsic component of synovial fluid) [42, 43]. The main function of the synovial fluid is to distribute nutrients and gases to the upper regions of the cartilage tissue whilst providing lubrication at the interface between articulating surfaces in the joint to allow facilitation of movement [44].

7.2.2 Clinical Treatments and Challenges

Cartilage disease and traumas are a serious clinical problem that can lead to debilitating pain, swelling, inflammation, and can result in drastically impaired mobility. An ageing worldwide population is resulting in increased number of patients suffering from osteoarthritis, a leading cause of cartilage damage [45]. Additionally, tissue degeneration through disease or significant trauma results in steady degradation of the tissue with limited successful clinical interventions to prevent this long-term decline which eventually results in a total joint replacement (Fig. 7.3). The limited regenerative capacity of cartilage is due to the avascular nature of the tissue and low metabolic activity of the resident chondrocytes and is in marked contrast to other tissues such as bone and the liver which have remarkable healing properties in comparison [47, 48]. Current clinical interventions have limited success in regenerating or replacing articular cartilage with most therapies producing inferior fibrocartilage or only delaying the progression of tissue degeneration [49].

The most commonly used cartilage repair strategies are debridement and microfracture, autologous chondrocyte implantation (ACI), matrix-assisted autologous chondrocyte transplantation/implantation (MACT/MACI), autologous matrix-induced chondrogenesis (AMIC), and osteochondral auto- and allografts (Fig. 7.4) [50, 51]. Each of these approaches has major disadvantages such as the formation of inferior fibrocartilage, requirement of two surgeries, donor site morbidity, and limited prevention of continued tissue degeneration. Therefore, new approaches are required to address these unmet clinical challenges. TE has the potential for a breakthrough approach by synergistically incorporating chondrocytes/stem cells, biomaterials, 3D scaffolds, stimulatory growth factors, mechanical stimulation, and bioreactors for the fabrication of tissue constructs. If this approach is successful then we will approach our final goal of the predictable regeneration of articular cartilage [52–57].

7.3 Engineering Hydrogels for Cartilage Tissue Engineering

Hydrogels are three-dimensional (3D) insoluble crosslinked polymer networks able to retain a large amount of water and biological fluids (between 10% and 200%) in their swollen state [58–61]. The 3D network of the hydrogel (Fig. 7.5) is maintained by polymer crosslinking among the polymer chains, with water taken up and contained within the polymeric structure [62, 63].

Hydrogels have a range of useful characteristics such as their hydrophilic capacity to absorb a volume of water that is significantly beyond their weight when dehydrated [64], potential printability (requires engineering) and ability to create a self-supporting 3D structure to suit various biological engineering needs (cell encapsulation, cell support and biodegradability) [65], uniform cell distribution, as well as providing both chemical and biological signalling [66]. The ability of the 3D

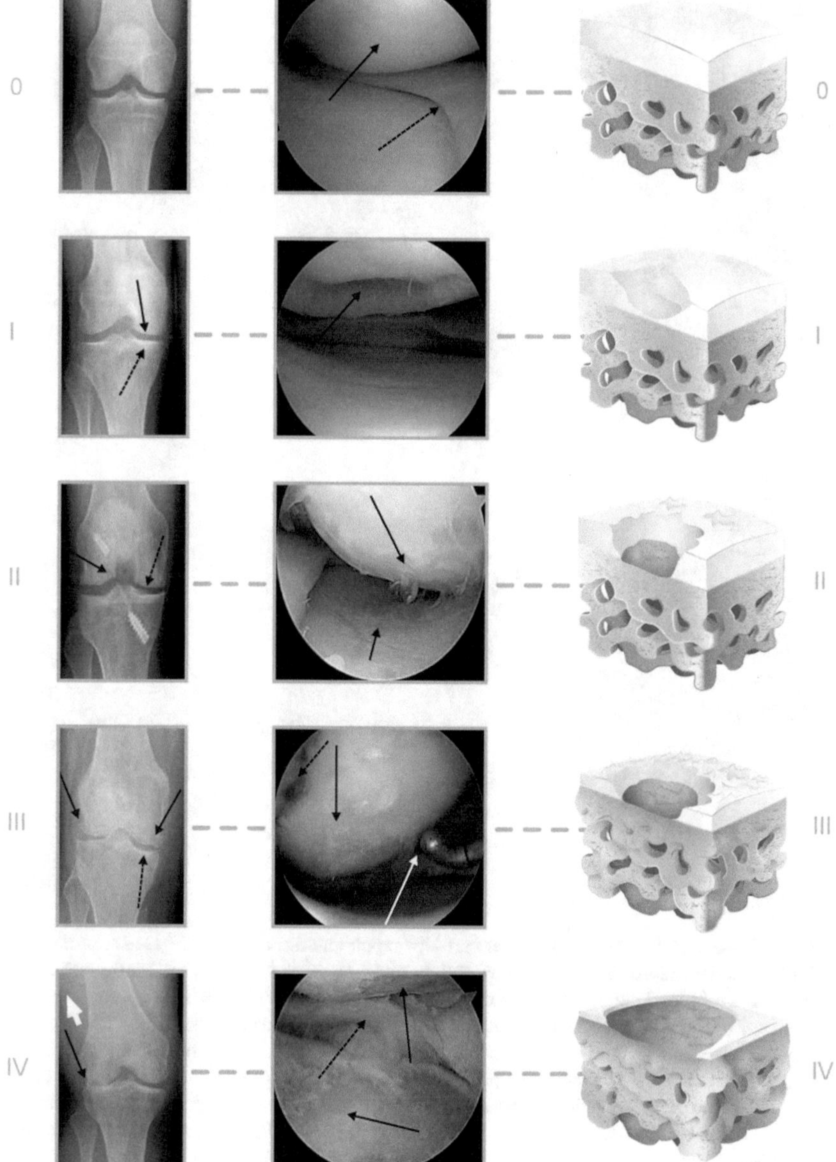

Fig. 7.3 Stages of articular cartilage damage. (**0**) Healthy cartilage. (**I**) Softening, swelling and fissuring of the cartilage. (**II**) Lesions <50% of cartilage depth. (**III**) Lesions >50% of cartilage depth but not through subchondral bone layer. (**IV**) Severely abnormal with penetration through the subchondral plate. (Reproduced with permission from the International Cartilage Regeneration & Joint Preservation Society (ICRS) [46])

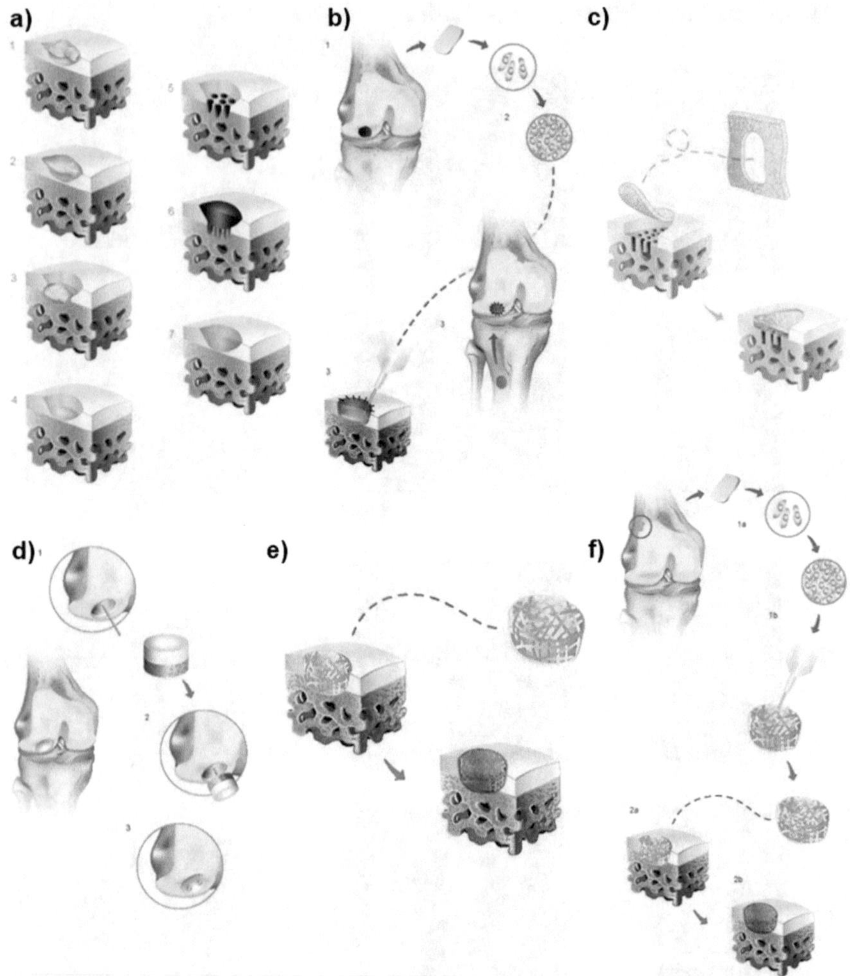

Fig. 7.4 Current techniques for articular cartilage repair and future scaffold-based approaches. (**a**) Debridement and microfracture; [1–4] debridement of damaged tissue, [5] microfracture into the subchondral bone, [6] blood clot formation containing mesenchymal stem cells (MSCs), and [7] formation of inferior fibrocartilage. (**b**) ACI; [1] cartilage biopsy from a non-load bearing site, [2] in vitro expansion of chondrocytes to obtain sufficient numbers for implantation, [3] a second operation cleans the damaged area and is covered with the periosteal flap and the expanded chondrocytes are injected under this or the expanded chondrocytes are seeded into a matrix (e.g. collagen membrane) and fixed with fibrin glue into the defect area (MACI/MACT). (**c**) AMIC; a microfracture is created and a membrane placed over the top and fixed with a fibrin glue or sutures. The membrane protects the clot formation and allows chondrogenic differentiation of the MSCs in the blood to aid the regeneration process. (**d**) Osteochondral auto/allograft transplantation; cylindrical plug of fresh cartilage and subchondral bone is taken from the patient or a cadaver donor, [1] a guide pin is placed perpendicular to the joint surface in the defect, [2] implantation of the graft plug, and [3] securing the plugs with the insertion of screws. (**e**) Acellular scaffold only requiring a single surgery and in vivo recruitment and stimulation of cells for tissue regeneration. (**f**) Cellular-based scaffold involving two surgeries of initial cartilage biopsy, chondrocyte expansion, and seeding into the scaffold before implantation into the defect. (Adapted with permission from the ICRS [46])

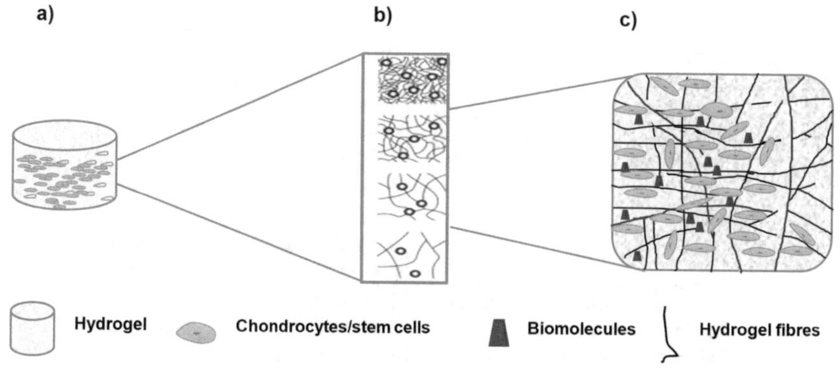

Fig. 7.5 Engineering 3D hydrogels for cartilage application through (**a**) cell-encapsulation, (**b**) optimising crosslinking density, and (**c**) functionalising with biomolecules (e.g. RGD) to guide cell behaviour

network to resist dissolution in water is related to the presence of the crosslinked polymer molecules which will determine the quality and stability of the hydrogel. Due to these properties and characteristics, a wide range of different hydrogels have been used for ECM formation and cartilage regeneration applications. Currently, natural based hydrogels represent the most relevant group of materials to produce constructs for cartilage applications (Table 7.1). Depending on the source, they can mimic the ECM structure and composition of articular cartilage presenting relevant swelling, degradation, mechanical and lubrication characteristics. Natural based hydrogel materials such as alginate [85], agarose [86], gelatin [87], collagen [83, 84, 88], hyaluronic acid (HA) [89], chitosan [90], and interpenetrating network/hybrid systems have been extensively explored for cartilage applications [19].

Although pure natural hydrogels have been widely explored in cartilage applications, they often have significant limitations relating to the suitability of their biomechanics, degradation profiles, and cell-material biological interactions. Subsequently, significant research into new hydrogel systems that have tuneable mechanical and degradation properties, provide specific control over cellular behaviour, allow incorporation of novel materials, and display suitable material processing properties for different fabrication routes such as 3D printing have been developed [91–94]. The incorporation of multiple materials, hybrid hydrogels, to produce a system with a mixture of two or more of distinct classes of molecules (e.g. a blend of natural and synthetic hydrogels) or different structures (e.g. blend of organic polymer hydrogels and inorganic particles/metals or synthetic nanoparticles) have been developed to improve on single material-based systems (Fig. 7.6).

The use of photocrosslinkable hybrid hydrogels is a burgeoning research field especially within tissue engineering. Photocurable hydrogels are particularly relevant for biomedical applications due to the advantage of being able to encapsulate cells, fast gelation time, tuneable mechanical properties and crosslinking density, and the cytocompatible processing conditions that allow their in situ crosslinking

Table 7.1 Photocrosslinkable natural polymers used in cartilage tissue engineering

Polymer	Crosslinking conditions	Biological properties	References
Alginate	Methacrylate anhydride, photoinitiators (VA-086, Irgacure 2959) and 254–365 nm exposure wavelength. Simple preparation.	Chondrocytes showed ECM formation with ~80% similarity to native cartilage. >85% viability	[67–69]
		Stem cells/ MSCs/ hASCs: could be regulated via the introduction of soluble factors and biophysical cues in 3D cell culture systems	
Agarose	Physical crosslinked (thermal) can be functionalised with methacrylate groups	Chondrocytes: no results reported (for chondrocyte cells encapsulated in photocurable agarose).	[70–72]
		hBMSCs: cell viability in agarose scaffolds decreased from 75% at day 3 to 40% at day 90.	
		Fibrochondrocytes: 80% cell viability and maintains chondrocyte phenotype	
Gelatin	Methacrylate and acrylic anhydride, photoinitiators (VA-086, Irgacure 2959) and 254–365 nm exposure wavelength. Simple preparation.	Chondrocytes showed ECM formation similar to native cartilage. >85–97% viability, cell morphology could be controlled by the stiffness of hydrogels.	[73–76]
		Stem cells/ MSCs/ hASCs: could be regulated via the introduction of soluble factors and biophysical cues in 3D cell culture systems	
Hyaluronic acid	Glucuronic acid carboxylic acid, the primary and secondary hydroxyl groups, and the N -acetyl group (following deamidation) thiols, methacrylates, tyramines Photoinitiators (VA-086, Irgacure 2959) and 254–365 nm exposure wavelength	Chondrocytes, epithelial cells and murine fibroblasts cells For cartilage engineered tissue, ECM formation reached about 80% of those found in native cartilage. >85% viability	[74]
		Promoted the retention of chondrocyte phenotype and matrix synthesis of encapsulated porcine chondrocytes and enhanced cell infiltration and tissue repair in rabbit osteochondral defect model at 2-week post-surgery	
		Stem cells/ MSCs/ hASCs: could be regulated via the introduction of soluble factors and biophysical cues in 3D cell culture systems	

(continued)

Table 7.1 (continued)

Polymer	Crosslinking conditions	Biological properties	References
Chitosan (CS)	Vinylated CS Macromers, Azido-functionalised CS azido, and (methyl)acrylated CS	Structurally analogous to cartilage glycosaminoglycans, Biocompatible, showed low cytotoxicity at various concentrations for chondrocytes, cell viability between 82% and 96%	[77, 78]
Dextran	Methacrylate groups (glycidyl methacrylate) introduced in dextran to prepare photo-crosslinkable dextran using Irgacure 2959 as a photoinitiator and 254–365 nm exposure wavelength	Chondrocytes, endothelial cells and stem cells, the dextran-based IPN hydrogel provides cell-adhesive and enzymatically degradable properties Fibroblasts ~90% cell viability.	[79, 80]
Gellan gum	Methacrylic anhydride, the aqueous solution of methacrylated gellan is added with 0.08 mg/mL calcium chloride and 0.5% (w/v) Irgacure 2959	Human nasal chondrocytes: The gellan gum supports is their use as cells encapsulating agents to be used in cartilage regeneration approaches Exhibits good cell viability and chondrocyte morphology.	[81, 82]
Collagen	Collagen I: dual crosslinking with visible light, Rose Bengal, and chemical crosslinking using 1-ethyl-3-(3-dimethylaminopropyl) carbodiimide and N-hydroxysuccinimide.	Chondrocytes had ≥95% viability. Dual-crosslinked gels showed greater resistance to collagenase digestion.	[83]
	Collagen II: amidation reaction between amino groups on the collagen lysine and methacrylic anhydride resulting in collagen methacrylamide. Irgracure 2959 used as photoinitiator and exposed to UV (8 W cm^{-2}) for 30 s. Triple helical structure of collagen was preserved after functionalisation.	BMSCs differentiated into a chondrogenic phenotype with lacuna-like cartilage formation. High cell viability. Increase in chondrogenic gene expression (AGG, SOX9, COL II, COL X). In vivo subcutaneous implantation in a mouse model demonstrated the secretion and deposition of cartilage specific matrix.	[84]

into the defect site during a surgical procedure [100–102]. The use of vat photopolymerisation and light curing processes in additive manufacturing also enables the fabrication of high resolution and complex structures that are not possible through conventional methods [17, 103]. This approach facilities the development of more biomimetic structures that can recapitulate the complexities of native tissues.

Typically, polymers are unable to initiate a photopolymerisation reaction alone and can require both functionalisation with photocrosslinkable groups and the addition of light sensitive molecules, called photoinitiators. The photoinitiators typically used within tissue engineering are normally excitable within the ultraviolet (UV)

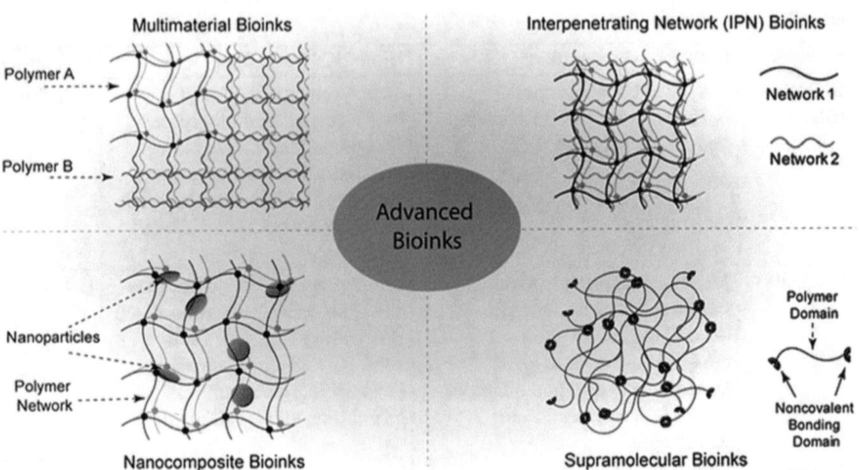

Fig. 7.6 Hybrid hydrogel network systems [95]. Multi-material bioink networks are two different component polymers or more crosslinked together. Interpenetrating networks (IPN) is a 3D structure of two or more polymeric networks, which are partially or fully interlaced at the molecular scale but not covalently bonded, or a network of embedded linear polymers entrapped within the original hydrogel (semi-IPN) [96, 97]. Nanocomposites bioink network can be produced by adding nanoparticles to the polymeric hydrogels [98]. Supramolecular bioink network is composed of short repeating units with functional groups that can interact non-covalently with other functional units, forming large, polymer-like entanglements [99]

(eg. 1-[4-(2-hydroxyethoxy)-phenyl]-2-hydroxy-2-methyl-1-propane-1-one, Irgacure 2959, and 2,2′-azobis[2-methyl-N-(2-hydroxyethyl)propionamide, VA-086], UV-visible (e.g. lithium phenyl-2,4,6-trimethylbenzoylphosphinate (LAP), and diphenyl(2,4,6-trimethylbenzoyl) phosphine oxide, TPO), and visible light range (e.g. ruthenium, Bengal rose, riboflavin, eosin-Y, and camphorquinone)). Upon irradiation, an initiation step begins and the photoinitiator undergoes an excitation mechanism which results in the formation of highly reactive species (free radicals). The next step is propagation, free radicals attach at specific sites on the monomer/oligomer structure (unsaturation sites) starting a chain reaction that creates a permanent crosslinked network. Finally, a termination step results in the completion of the polymer chain and crosslinking with free radical extinction. The degree of photopolymerisation (curing percentage) of the resulted hydrogels is highly effected by the functionalisation process and polymer solution concentration, light wavelength and intensity, photoinitiator type and concentration, and curing time [104]. The number of parameters to control also provides an opportunity to highly engineer the resulting hydrogels properties and when used in conjunction with bioprinting processes can result in tuneable and biomimetic structures.

Photocrosslinking is an advantageous process, but specific issues need to be addressed especially in biomedical applications. The chemical processes used can generate toxic by-products or residues that can result in cytocompatibility issues in the hydrogel [105]. Thus, the hydrogel is usually subjected to a purification step prior to its application, for example, the modified macromer may require dialysis prior to use. Furthermore, the photoinitiators themselves can be toxic so specific care is required when designing the photocuring system such as using a low concentration of photoinitiator. Finally, the wavelength and intensity of light can induce genetic damage within the cells resulting in reduced viability and function, thus this requires careful consideration [106–110].

Crosslinking mechanism, biofunctionalisation, controlled delivery of biochemical factors and hydrogel physical characteristics are the most crucial parameters that determine hydrogel suitability for the target application.

7.3.1 Crosslinking Mechanisms of Hydrogels

The crosslinking mechanisms of hydrogels are broadly classified into two categories (Fig. 7.7): chemically crosslinking (permeant crosslinking) and physical crosslinking (reversible crosslinking) [112]. Chemically crosslinking hydrogels are formed by covalent bonding that creates a permanent 3D polymer network [113]. Physical crosslinking hydrogels are produced through the entanglement mechanism of the polymer chains [114]. For the fabrication of both scaffolds and cell-laden constructs, chemical crosslinking is the most suitable mechanism [105, 115].

Crosslinking reactions can be induced by using a variety of stimuli such as light, pH stimulation, ionic exchange, enzymes, and temperature [105]. The crosslinking density strongly determines the characteristics of the hydrogels, such as mechanical, degradation, porosity, pore size, swelling and biological properties [116, 117].

Light initiated crosslinking reactions correspond to the most common approach, allowing the fabrication of complex structures under relatively mild conditions. Two different reaction mechanisms are being considered: free radical polymerisation and click chemistry.

Free radical photopolymerisation is a chain growth reaction started by the absorption of light by photoinitiators. The photoinitiator concentration determines

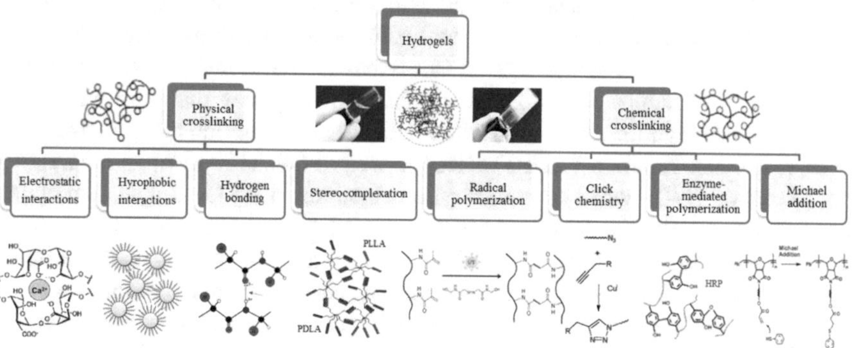

Fig. 7.7 Crosslinking mechanisms using in physical and chemical gelation of hydrogels [111]

the polymerisation kinetics and consequently the properties of the polymerised structures. Radical photopolymerisable hydrogels can be functionalised with cell adhesive moieties and degradation sites in a relatively easy and reproducible manner [118]. Major limitations are associated to the relatively poor control over the crosslinking kinetics, the presence of unreacted double bonds which can react with biological substances, and the generation of heterogeneities due to the random chain reaction process. As an alternative, orthogonal click reactions (e.g. thiol-norbornene) proceed under mild reaction conditions with higher efficiency and faster curing kinetics compared to free radical photopolymerisation [119]. This step-growth reaction allows high spatio-temporal control and the fabrication of 3D networks with minimal defects [120].

7.3.2 *Biofunctionalisation of Hydrogels*

Hydrogels for cartilage tissue engineering must be designed to closely capture the composition, architecture, biophysical properties and biochemical cues of the native tissue. This is essential to provide a cell-instructive 3D microenvironment that guides the fate of embedded cells and morphogenesis. In the case of acellular networks, hydrogel characteristics such as mechanical properties and porosity must ensure adequate load transfer and eventually promote cell recruitment. Multiple approaches have been explored to develop photocrosslinkable hydrogels that perform fundamental functions of the ECM, including the (i) material selection and hydrogel structure, (ii) modulation of hydrogel viscoelasticity, (iii) bioconjugation of bioactive peptides, and (iv) tethering of biochemical factors [121].

7.3.2.1 Hydrogel Composition and Structure

Important design criteria for hydrogels include appropriate mechanical properties, ensuring nutrient/waste transport and maintenance of cell viability. Several studies have modulated the mechanical properties of hydrogels by varying the macromer concentration or crosslinker type/content using a single polymer network. These works demonstrated that hydrogel stiffness has a tremendous impact on matrix deposition with highly crosslinked networks restricting new tissue formation and eventually leading to matrix calcification [122, 123]. In addition to the matrix stiffness, it was also reported that the molecular weight (50–1100 kDa) and macromer concentration (2–20 wt%) of methacrylated hyaluronic acid hydrogels determine the amount and distribution of new ECM via modulation of the gel structure (e.g., compressive modulus, swelling ratio) [124]. Another important consideration of hydrogels is to balance the mechanical integrity and the degradation rate. Gel networks should support tissue growth as the hydrogel degrades, maintaining the structural integrity of the new ECM. One interesting approach to control the degradation rate of hydrogels for cartilage applications involves the synthesis of photocrosslinkable macromers bearing hydrolytically labile linkages. This strategy was explored to design hydrolytically degradable hydrogels via thiol-ene photopolymerisation of norbornene-modified Poly(ethylene glycol) (PEG) containing caprolactone segments with hydrolytically labile ester linkages and PEG dithiol crosslinker in the presence of a photoinitiator [125]. Although hydrogels based on a single polymer are relatively simple to design, their limited mechanical properties and potential reduced cell viability due to high crosslink density constitute important concerns. To tackle these limitations, more complex hydrogels have been developed based on the combination of multiple polymers. Depending on the selected materials and crosslinking chemistries, different hydrogel structures have been designed, including interpenetrating networks (IPNs) [126]. semi-IPNs [127], double networks [128] and guest-host networks [129]. Wei et al. [130] synthesised a two-component guest-host hydrogel of adamantane-functionalised hyaluronan (HA) as guest polymers and monoacrylated β-cyclodextrin as host monomers, rapidly crosslinked through UV photopolymerisation. Hydrogels displayed self-healing properties, supported robust chondrogenesis of embedded human MSCs and stimulated cartilage formation in a rat model. In addition to the dynamic nature of the network and/or superior mechanical properties often achieved by more complex hydrogel systems, these gels can also address key compositional and biological properties of the native tissue by the incorporation of major components of the cartilage ECM [131]. One strategy consists of the chemical incorporation of ECM components bearing a photocrosslinkable moiety into the hydrogel network via photopolymerisation. This strategy was explored by Wang et al. [132] to evaluate the influence of methacrylated ECM molecules (chondroitin sulfate, heparan sulfate and hyaluronic acid) on the in vivo cartilage formation using ECM-containing hydrogels with tuneable stiffness and controllable biochemical composition. Methacrylated ECM molecules were incorporated into the network of PEG dimethacrylate hydrogels through photopolymerisation in the presence of adipose derived stem cells and neonatal chondrocytes. Cellular hydrogels with varying matrix stiffness and biochemical cues showed increased collagen type II deposition in both chondroitin sulfate- and hyaluronic acid-containing hydrogels, while heparan sulfate-containing hydrogels promoted a fibrocartilage phenotype and failed to retain newly deposited matrix. This study revealed the importance of biochemical composition of hydrogels and matrix stiffness on the phenotype of neocartilage formation.

7.3.2.2 Hydrogel Viscoelasticity

Traditional hydrogels obtained by photopolymerisation reactions are made of strong covalent bonds, leading to almost purely elastic gel networks. In contrast to native cartilage tissue, which is viscoelastic and exhibit partial stress relaxation when a constant strain is applied, chemical hydrogels store cellular forces and resist deformation. Purely elastic hydrogels with embedded chondrocytes were found to limit cell proliferation and restrict cartilage matrix deposition to the pericellular space, probably due to the elastic nature of hydrogels [133]. Recent findings have shown that viscoelastic hydrogels, in which the stress is relaxed over time, support the spreading, proliferation and differentiation of embedded mesenchymal stem cells in 3D culture without requiring matrix degradation [134]. It has also been reported that cell-laden hydrogels exhibiting faster stress relaxation support significantly more new bone growth in vivo when compared to slow stress-relaxing gel networks [135]. As stress relaxation is suggested as a key design parameter of hydrogels to both better recapitulate the viscoelastic behaviour of natural ECM and perform important biological behaviours, efforts have been focused on the design of viscoelastic hydrogels with tuneable stress relaxation for cartilage tissue engineering. In a recent work, Lee et al. [136], developed a series of viscoelastic alginate hydrogels exhibiting tuneable stress relaxation to evaluate the effects of viscoelasticity on embedded chondrocytes in 3D culture. Stress relaxation of calcium-crosslinked hydrogels was controlled by varying the molecular weight of alginate and covalent coupling of short PEG spacers. It was found that hydrogels with faster stress relaxation supported the formation of increased cartilage matrix with deposition of both collagen and GAGs (Fig. 7.8), while slower relaxing gel networks led to the up-regulation of genes associated to cartilage degradation and cell death. An alternative approach to engineer hydrogels with tuneable viscoelastic properties consists on the formation of covalent adaptable networks with the ability to reorganise the network connectivity to dissipate local stress. As an example, hydrazone crosslinked PEG hydrogels were synthesised as viscoelastic gels for cartilage applications [137]. After 4 weeks of culture, porcine chondrocytes embedded within hydrogels exhibiting average relaxation times of 3 days secreted an interconnected articular cartilage-specific matrix with increased deposition of collagen (e.g., collagen type II) and sulfated glycosaminoglycans (e.g., aggrecan) compared to predominantly elastic hydrogels with slow average relaxation times (~1 month). The authors suggested that a balance between slow and fast relaxing crosslinks is essential to preserve gel network integrity and support the formation of high-quality neocartilaginous tissue.

7.3.2.3 Bioconjugation of Bioactive Peptides

Articular cartilage is an intricate meshwork rich in collagen and proteoglycans. Rather than a static milieu, the ECM is a highly dynamic microenvironment that establishes reciprocal interactions with neighbouring cells. Cartilage ECM components provide biochemical motifs for cell adhesion and degradation sites for cell-mediated matrix degradation, which are essential to allow cell adhesion, proliferation, migration and tissue formation. As most cell types require adhesion sites to survive and perform their functions, several bioconjugation strategies have been explored for the functionalisation of hydrogels with bioactive peptides that impart cell adhesion domains [138]. The fibronectin-derived adhesion peptide RGD (arginine-glycine-aspartic acid) is often bond to the hydrogel network to promote integrin-mediated cell binding and allow cell attachment and spreading [126, 139]. Other sequences derived from collagen and decorin have also been immobilised

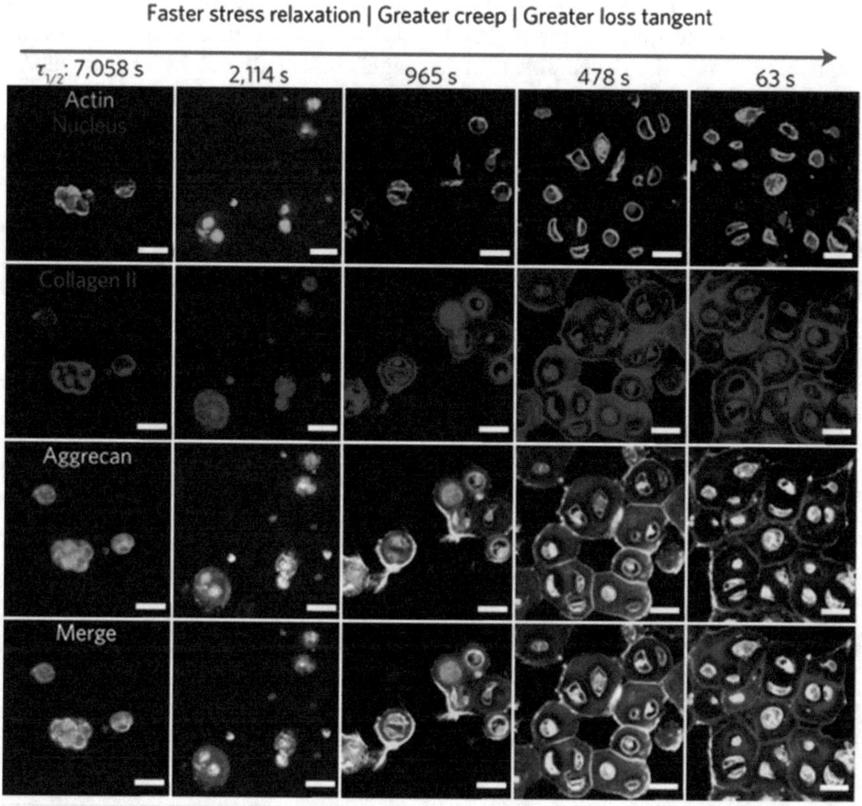

Fig. 7.8 Influence of hydrogel stress relaxation on cartilage matrix production and formation. Immunohistochemical staining of chondrocytes embedded within 3 kPa alginate hydrogels for 21 days (scale bar: 25 μm) [136]

into hydrogels with beneficial outcomes regarding chondrogenic differentiation of MSCs and matrix retention [140, 141]. Since chondrocytes in native ECM display a round morphology, the benefit of RGD functionalisation in chondrocytes is still controversial. For example, chondrogenesis of bone marrow stromal cells embedded within RGD-functionalised alginate hydrogels was inhibited in response to TGF-β1 and dexamethasone regarding gene expression and matrix synthesis [142]. Another work showed that photocrosslinked RGD-modified PEG hydrogels enhanced cartilage-specific gene expression and matrix synthesis by bovine chondrocytes, but only in the presence of dynamic mechanical stimulation [143]. In fact, most of the hydrogel systems used for chondrocyte encapsulation did not require cell-adhesion domains to support tissue formation [127, 136]. Overall, this data demonstrates that the immobilisation of cell-adhesion domains in hydrogels perform a key role in terms of cell phenotype and chondrogenesis, further studies are necessary to elucidate the cell-specific influence of peptide ligands.

Natural ECM is continuously remodelled by cell-secreted enzymes, which is required to create space in the matrix for cell migration, proliferation and tissue deposition. Thus, the development of hydrogels susceptible to hydrolytic and/or cell-driven degradation is becoming increasingly recognised as a key feature to improve the ECM biomimicry [144]. Hydrolytic degradation of hydrogels is commonly obtained by hydrolysis of ester linkages, leading to changes in the overall network properties. While enzymatic degradation depends on the cell type and levels of circulating enzymes, hydrolytic degradation is spontaneous and occurs via bulk and/or surface erosion mechanisms. Cell-driven hydrogel degradation via cleavage of protease-sensitive peptides occurs in a more localised manner, at the pericellular space, better recapitulating the remodelling of natural ECM. Hydrogels based on animal polymers such as fibrin, collagen and hyaluronic acid are naturally degraded, whereas some plant-based polymers (e.g., alginate, pectin) and synthetic polymers are often modified to make them susceptible to enzymatic degradation [138]. This has been accomplished by the incorporation of peptide sequences recognised by specific proteases. These peptides can be grafted into the polymer backbone or used as crosslinker agents with a dual function – allow gel formation and promote matrix degradation. As an example, MMP7-degradable hydrogels based on recombinant Streptococcal collagen-like 2 (Scl2) proteins and functionalised with peptides that bind to hyaluronic acid and chondroitin sulphate were synthesised for cartilage tissue engineering [145]. Cell-degradable hydrogels with embedded hMSCs and functionalised with GAG-binding peptides significantly enhanced chondrogenic differentiation and gene expression of COL2A1, ACAN, and SOX9, compared to non-functionalised hydrogels.

7.3.2.4 Tethering and Controlled Delivery of Biochemical Factors

Cartilage formation is regulated by a coordinated spatiotemporal delivery of biochemical factors that interact with cells to induce chondrogenesis. To this end, a major role of ECM is to serve as a storage depot for growth factors, regulating their spatiotemporal and tissue-specific presentation to the cells. Its ability to locally bind, store, and release growth factors is essential to regulate their bioavailability, bioactivity and stability in order to elicit a biological response [146]. The binding of growth factors to the ECM, for example, via electrostatic interactions to heparan sulfate, is fundamental to enhance their activity in the vicinity of cells, prolong their action and protect them from degradation [147].

To recreate the dynamic presentation of bioactive cues available to the cells in native cartilage, hydrogels have been designed with the ability to sequester and/or deliver biochemical cues such as growth factors. This has been achieved through different strategies including the covalent immobilisation [148], affinity binding [149], and encapsulation within carrier vehicles [150]. Members of the transforming growth factor beta (TGF-β) superfamily such as TGF-β1 play a role in regulating chondrogenesis during development. To allow the immobilisation into the hydrogel network towards improved bioactivity and cell presentation, growth factors are usually modified with photoreactive groups. In one example, TGF-β1 was reacted with 2-iminothiolane, yielding thiolated TGF-β1 that can homogeneously bond throughout the gel network of non-degradable PEG hydrogels during photopolymerisation without impairing its bioactivity [148]. Thiolated TGF-β1 was covalently linked to the norbornene end groups of PEG hydrogels crosslinked by photoinitiated step-growth polymerisation using an MMP-degradable peptide (KCGPQGIWGQCK) as a crosslinker [151]. The ability of chondrocytes to promote in situ hydrogel degradation was firstly confirmed using a fluorogenic peptide sensor to determine the extent of MMP-sensitive sequence cleavage (Fig. 7.9). In this hydrogel, GAG and collagen deposition was found to be restricted to the pericellular space, which was attributed to the limited chondrocyte degradation of the

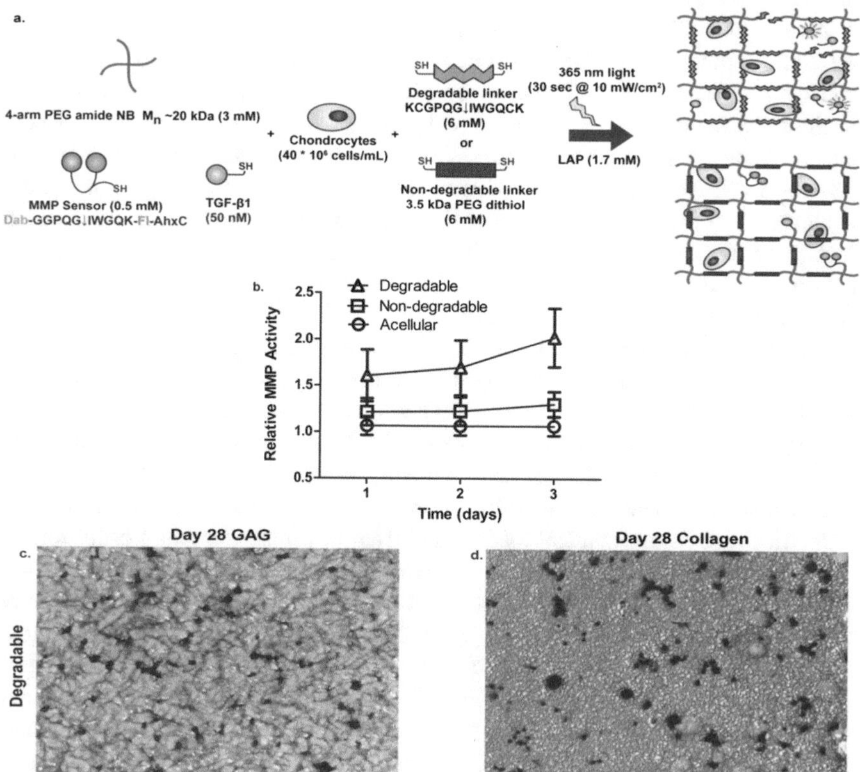

Fig. 7.9 Effect of chondrocytes embedded within cell-degradable hydrogels. (**a**) Schematic illustration of cell-laden hydrogel formation with tethered MMP fluorescent sensor (Dab-GGPQG↓IWGQK-Fl-AhxC), and TGF-β1; hydrogels are crosslinked using either an MMP-degradable peptide sequence (KCGPQG↓IWGQCK) or non-degradable (3.5 kDa PEG dithiol) linker. (**b**) Determination of in situ cleavage of fluorescent sensor by chondrocytes. (**c**) GAG staining of sections from cell-laden hydrogels after 28 days of culture: nuclei stained black and GAGs stained red. (**d**) Collagen staining of sections obtained from cell-laden hydrogels at day 28: nuclei stained black and collagen stained blue (scale bars: 100 μm) [151]

gel network to allow diffuse matrix production. When hydrogels were used to co-encapsulate chondrocytes and hMSCs, a higher degradation rate was observed, along with increased GAG and collagen deposition at 14 days of culture compared to non-degradable hydrogels. This data highlights the importance of cell-degradable motifs and localised presentation of biochemical cues in photocrosslinked hydrogels to support the formation of cartilage specific ECM.

7.4 Conclusions and Future Perspectives

Naturally derived hydrogels are suitable materials for cartilage applications. They are biocompatible and biodegradable with a chemical and physical structure that resembles the ECM structure of natural tissues. Different light-mediated photopolymerisation strategies have been explored for the fabrication of scaffolds with micro- and nano-scale resolutions. These strategies proceed under biocompatible conditions in the presence of cells and biochemical signals allowing the fabrication of cell-laden bioconstructs. In order to improve the biological performance of both scaffolds and cell-laden constructs different biofunctionalisation strategies have been explored. Many naturally derived hydrogels degrade through enzymatic and/or hydrolytic mechanisms, adding another layer of complexity to control their degradation rate. This is an important limitation in terms of the design of suitable constructs for cartilage tissue engineering.

Another critical challenge in cartilage tissue engineering is the integration and engineering of hydrogels that are specific for each region of the osteochondral tissue. The structure and properties of constructs for osteochondral applications must address the characteristics of the two comprising tissues, articular cartilage and subchondral bone, which still represents a major research challenge. The section of such constructs related to the cartilage zone must present adequate mechanical properties to resist mechanical loading and friction whilst promoting ECM formation and chondrogenic expression of mesenchymal stem cells or chondrocytes, inhibiting hypertrophic differentiation and mineralisation of chondrocytes in the upper regions of the tissue. Contrary, the section of the constructs corresponding to subchondral bone must promote the formation of a blood vessel network, stimulate osteoblast proliferation, and osteogenic differentiation of mesenchymal stem cells whilst remaining biomechanically suitable until successful tissue regeneration is complete. How to engineer hydrogels to promote or inhibit specific cellular behavioural pathways is a major challenge, for example, a osteochondral hydrogel must promote hypertrophic chondrocytes in the calcified zone and flatter and aligned chondrocytes in the superficial surface. This will demand the development of multi-material and multi-scale hydrogels that are specifically biofunctionalised to promote the complex behaviour and organisation of chondrocyte cells. Furthermore, top-down fabrication techniques will need to improve their resolution and multi-material processing capabilities thus driving demand for hybrid biomanufacturing systems that incorporate multiple fabrication technologies into a single system. Whilst bottom-up fabrication processes will require further understanding of the fundamental developmental biology of osteochondral tissue.

Acknowledgements The authors wish to acknowledge the support of the Government of Iraq for supporting a PhD through a grant provided by the Higher Committee for Development Education Iraq (HCED) and the funding provided by the Engineering and Physical Sciences Research Council (EPSRC) and Medical Research Council (MRC) Centre for Doctoral Training in Regenerative Medicine (EP/L014904/1).

References

1. R. Langer, J.P. Vacanti, Tissue engineering. Science (New York, NY) **260**(5110), 920 (1993)
2. M.M. Stevens, Toxicology: testing in the third dimension. Nat Nanotechnol **4**(6), 342 (2009)
3. E.S. Place, J.H. George, C.K. Williams, M.M. Stevens, Synthetic polymer scaffolds for tissue engineering. Chem Soc Rev **38**(4), 1139–1151 (2009)
4. G. Jell, R. Swain, M.M. Stevens, Raman spectroscopy: a tool for tissue engineering, in *Emerging Raman Applications and Techniques in Biomedical and Pharmaceutical Fields*, (Springer, Heidelberg, 2010), pp. 419–437
5. E.S. Place, N.D. Evans, M.M. Stevens, Complexity in biomaterials for tissue engineering. Nat. Mater. **8**, 457 (2009)
6. A.R. Verissimo, K.J.D.P. Nakayama, Scaffold-free biofabrication, in *3D Printing and Biofabrication*, (Springer, 2018), pp. 431–450
7. G.D. DuRaine, W.E. Brown, J.C. Hu, K.A. Athanasiou, Emergence of scaffold-free approaches for tissue engineering musculoskeletal cartilages. Ann Biomed Eng **43**(3), 543–554 (2015)
8. D.J. Huey, J.C. Hu, K.A. Athanasiou, Unlike bone, cartilage regeneration remains elusive. Science **338**(6109), 917–921 (2012)
9. Y. Jung, H. Ji, Z. Chen, H. Fai Chan, L. Atchison, B. Klitzman, et al., Scaffold-free, human mesenchymal stem cell-based tissue engineered blood vessels. Scientific reports **5**, 15116 (2015)
10. A. Ovsianikov, A. Khademhosseini, V. Mironov, The synergy of scaffold-based and scaffold-free tissue engineering strategies. Trends Biotechnol. **36**(4), 348–357 (2018)
11. C.A. DeForest, K.S. Anseth, Advances in bioactive hydrogels to probe and direct cell fate. Ann Rev. Chem. Biomol. Eng. **3**, 421–444 (2012)
12. M.P. Lutolf, Biomaterials: spotlight on hydrogels. Nat. Mater. **8**(6), 451–453 (2009)
13. G. Camci-Unal, N. Annabi, M.R. Dokmeci, R. Liao, A. Khademhosseini, Hydrogels for cardiac tissue engineering. NPG Asia Mater. **6**, e99 (2014)
14. A.S. Hoffman, Hydrogels for biomedical applications. Adv. Drug Deliv. Rev. **54**(1), 3–12 (2002)
15. H.L. Lim, Y. Hwang, M. Kar, S. Varghese, Smart hydrogels as functional biomimetic systems. Biomater. Sci. **2**(5), 603–618 (2014)
16. M.P. Lutolf, J.L. Lauer-Fields, H.G. Schmoekel, A.T. Metters, F.E. Weber, G.B. Fields, et al., Synthetic matrix metalloproteinase-sensitive hydrogels for the conduction of tissue regeneration: Engineering cell-invasion characteristics. PNAS **100**(9), 5413–5418 (2003)
17. R.F. Pereira, P.J. Bártolo, 3D bioprinting of photocrosslinkable hydrogel constructs. J Appl Polym Sci **132**(48) (2015)
18. J. Martel-Pelletier, A.J. Barr, F.M. Cicuttini, P.G. Conaghan, C. Cooper, M.B. Goldring, et al., Osteoarthritis. Nat. Rev. Dis. Primers. **2**, 16072 (2016)

19. I.L. Kim, R.L. Mauck, J.A. Burdick, Hydrogel design for cartilage tissue engineering: a case study with hyaluronic acid. Biomaterials **32**(34), 8771–8782 (2011)
20. P.R. van Weeren, General anatomy and physiology of joints, in *Joint disease in the horse*, (Elsevier, Amsterdam, Netherlands, 2016), pp. 1–24
21. J. Perera, P. Gikas, G. Bentley, The present state of treatments for articular cartilage defects in the knee. Ann R Coll Surg Engl **94**(6), 381–387 (2012)
22. A. Armiento, M. Stoddart, M. Alini, D. Eglin, Biomaterials for articular cartilage tissue engineering: Learning from biology. Acta Biomater **65**, 1–20 (2018)
23. D.F. Duarte Campos, W. Drescher, B. Rath, M. Tingart, H. Fischer, Supporting biomaterials for articular cartilage repair. Cartilage **3**(3), 205–221 (2012)
24. K.J. Jones, W.L. Sheppard, A. Arshi, B.B. Hinckel, S.L. Sherman, Articular cartilage lesion characteristic reporting Is highly variable in clinical outcomes studies of the knee. Cartilage **10**(3), 299–304 (2018). https://doi.org/10.1177/1947603518756464
25. S.P. Nukavarapu, D.L. Dorcemus, Osteochondral tissue engineering: Current strategies and challenges. Biotechnol. Adv. **31**(5), 706–721 (2013)
26. G.P. Huang, A. Molina, N. Tran, G. Collins, T.L. Arinzeh, Investigating cellulose derived glycosaminoglycan mimetic scaffolds for cartilage tissue engineering applications. J Tissue Eng Regen Med **12**(1), e592–e603 (2018)
27. T. Guo, J. Lembong, L.G. Zhang, J.P. Fisher, Three-dimensional printing articular cartilage: recapitulating the complexity of native tissue. Tissue Eng Part B Rev **23**(3), 225–236 (2017)
28. S. Camarero-Espinosa, B. Rothen-Rutishauser, E.J. Foster, C. Weder, Articular cartilage: From formation to tissue engineering. Biomater. Sci. **4**(5), 734–767 (2016)
29. R.J. Lories, F.P. Luyten, The bone-cartilage unit in osteoarthritis. Nat. Rev. Rheumatol. **7**(1), 43–49 (2011)
30. J.S. Temenoff, A.G. Mikos, Review: Tissue engineering for regeneration of articular cartilage. Biomaterials **21**(5), 431–440 (2000)
31. Z. Izadifar, X. Chen, W. Kulyk, Strategic design and fabrication of engineered scaffolds for articular cartilage repair. J Funct Biomater **3**(4), 799–838 (2012)
32. L. Zhang, J. Hu, K.A. Athanasiou, The role of tissue engineering in articular cartilage repair and regeneration. Crit. Rev. Biomed. Eng. **37**(1–2), 1–57 (2009)
33. B. Mollon, R. Kandel, J. Chahal, J. Theodoropoulos, The clinical status of cartilage tissue regeneration in humans. Osteoarthr. Cartil. **21**(12), 1824–1833 (2013)
34. A.J. Sophia Fox, A. Bedi, S.A. Rodeo, The basic science of articular cartilage: Structure, composition, and function. Sports Health **1**(6), 461–468 (2009)
35. E.B. Hunziker, Articular cartilage repair: Basic science and clinical progress. A review of the current status and prospects. Osteoarthr. Cartil. **10**(6), 432–463 (2002)
36. Z. Lin, C. Willers, J. Xu, M.H. Zheng, The chondrocyte: Biology and clinical application. Tissue Eng. **12**(7), 1971–1984 (2006)
37. WY.-w. Lee, B. Wang, Cartilage repair by mesenchymal stem cells: Clinical trial update and perspectives. J Orthop Trans **9**, 76–88 (2017)
38. L. Kjellen, U. Lindahl, Proteoglycans: Structures and interactions. Annu. Rev. Biochem. **60**, 443–475 (1991)
39. E. Sugawara, H. Nikaido, Properties of AdeABC and AdeIJK efflux systems of Acinetobacter baumannii compared with those of the AcrAB-TolC system of Escherichia coli. Antimicrob. Agents Chemother. **58**(12), 7250–7257 (2014)
40. Z. Abusara, M. Von Kossel, W. Herzog, In vivo dynamic deformation of articular cartilage in intact joints loaded by controlled muscular contractions. PLoS One **11**(1), e0147547 (2016)
41. S. Lusse, H. Claassen, T. Gehrke, J. Hassenpflug, M. Schunke, M. Heller, et al., Evaluation of water content by spatially resolved transverse relaxation times of human articular cartilage. Magn. Reson. Imaging **18**(4), 423–430 (2000)
42. J.A. Buckwalter, H.J. Mankin, Articular cartilage: Tissue design and chondrocyte-matrix interactions. Instr. Course Lect. **47**, 477–486 (1998)

43. C. Weiss, The physiologoy and pathology of hyaluronic acid in joints. Ups. J. Med. Sci. **82**(2), 95–96 (1977)
44. T.A. Schmidt, N.S. Gastelum, Q.T. Nguyen, B.L. Schumacher, R.L. Sah, Boundary lubrication of articular cartilage: Role of synovial fluid constituents. Arthritis Rheum. **56**(3), 882–891 (2007)
45. Y. Li, X. Wei, J. Zhou, L. Wei, The age-related changes in cartilage and osteoarthritis. BioMed Res Int **2013**, 916530 (2013)
46. International Cartilage Regeneration and Joint Preservation Society (ICRS). Other cartilaginous parts of the body. Available from: https://cartilage.org/patient/about-cartilage/welcome-to-our-joint/other-cartilaginous-parts-of-the-body/. Accessed 24 Apr 2019
47. D.J. Huey, J.C. Hu, K.A. Athanasiou, Unlike bone, cartilage regeneration remains elusive. Science (New York, NY) **338**(6109), 917–921 (2012)
48. C. Chung, J.A. Burdick, Engineering cartilage tissue. Adv. Drug Deliv. Rev. **60**(2), 243–262 (2008)
49. D. Correa, S.A. Lietman, Articular cartilage repair: Current needs, methods and research directions. Semin. Cell Dev. Biol. **62**, 67–77 (2017)
50. W. Swieszkowski, B.H. Tuan, K.J. Kurzydlowski, D.W. Hutmacher, Repair and regeneration of osteochondral defects in the articular joints. Biomol. Eng. **24**(5), 489–495 (2007)
51. S.N. Redman, S.F. Oldfield, C.W. Archer, Current strategies for articular cartilage repair. Eur. Cell. Mater. **9**, 23–32 (2005).; discussion 23-32
52. Y. Liu, G. Zhou, Y. Cao, Recent progress in cartilage tissue engineering—Our experience and future directions. Engineering **3**(1), 28–35 (2017)
53. G. Kalamegam, A. Memic, E. Budd, M. Abbas, A. Mobasheri, A comprehensive review of stem cells for cartilage regeneration in osteoarthritis. Adv. Exp. Med. Biol. **1089**, 23–36 (2018)
54. C. Loebel, J.A. Burdick, Engineering stem and stromal cell therapies for musculoskeletal tissue repair. Cell Stem Cell **22**(3), 325–339 (2018)
55. F.R. Maia, M.R. Carvalho, J.M. Oliveira, R.L. Reis, Tissue engineering strategies for osteochondral repair. Adv. Exp. Med. Biol. **1059**, 353–371 (2018)
56. M. Abbas, M. Alkaff, A. Jilani, H. Alsehli, L. Damiati, M. Kotb, et al., Combination of mesenchymal stem cells, cartilage pellet and bioscaffold supported cartilage regeneration of a full thickness articular surface defect in rabbits. J Tissue Eng Regen Med **15**(5), 661–671 (2018)
57. C. Vyas, G. Poologasundarampillai, J. Hoyland, P. Bartolo, 3D printing of biocomposites for osteochondral tissue engineering, in *Biomedical Composites*, ed. by L. Ambrosio, 2nd edn., (Woodhead Publishing, Sawston, Cambridge, UK, 2017), pp. 261–302
58. H. Omidian, K. Park, Hydrogels, in *Fundamentals and Applications of Controlled Release Drug Delivery*, ed. by J. Siepmann, R. A. Siegel, M. J. Rathbone, (Springer US, Boston, 2012), pp. 75–105
59. A. Barbetta, E. Barigelli, M. Dentini, Porous alginate hydrogels: Synthetic methods for tailoring the porous texture. Biomacromolecules **10**(8), 2328–2337 (2009)
60. J.H. Lee, G. Khang, J.W. Lee, H.B. Lee, Interaction of different types of cells on polymer surfaces with wettability gradient. J. Colloid Interface Sci. **205**(2), 323–330 (1998)
61. S. Utech, A review of hydrogel-based composites for biomedical applications: enhancement of hydrogel properties by addition of rigid inorganic fillers. J Mater Sci **51**(1), 271–310 (2016)
62. T. Billiet, M. Vandenhaute, J. Schelfhout, S. Van Vlierberghe, P. Dubruel, A review of trends and limitations in hydrogel-rapid prototyping for tissue engineering. Biomaterials **33**(26), 6020–6041 (2012)
63. M.C. Straccia, I. Romano, A. Oliva, G. Santagata, P. Laurienzo, Crosslinker effects on functional properties of alginate/N-succinylchitosan based hydrogels. Carbohydr. Polym. **108**, 321–330 (2014)
64. M. Liu, X. Zeng, C. Ma, H. Yi, Z. Ali, X. Mou, et al., Injectable hydrogels for cartilage and bone tissue engineering. Bone Res **5**, 17014 (2017)
65. Q.G. Wang, N. Hughes, S.H. Cartmell, N.J. Kuiper, The composition of hydrogels for cartilage tissue engineering can influence glycosaminoglycan profile. Eur. Cell. Mater. **19**, 86–95 (2010)

66. J. Malda, J. Visser, F.P. Melchels, T. Jungst, W.E. Hennink, W.J. Dhert, et al., 25th anniversary article: Engineering hydrogels for biofabrication. Adv Mater **25**(36), 5011–5028 (2013)
67. S. Fu, A. Thacker, D.M. Sperger, R.L. Boni, I.S. Buckner, S. Velankar, et al., Relevance of rheological properties of sodium alginate in solution to calcium alginate gel properties. AAPS PharmSciTech **12**(2), 453–460 (2011)
68. P. Stagnaro, I. Schizzi, R. Utzeri, E. Marsano, M. Castellano, Alginate-polymethacrylate hybrid hydrogels for potential osteochondral tissue regeneration. Carbohydr. Polym. **185**, 56–62 (2018)
69. A.D. Rouillard, C.M. Berglund, J.Y. Lee, W.J. Polacheck, Y. Tsui, L.J. Bonassar, et al., Methods for photocrosslinking alginate hydrogel scaffolds with high cell viability. Tissue Eng. Part C Methods **17**(2), 173–179 (2011)
70. I.D. Paepe, H. Declercq, M. Cornelissen, E. Schacht, Novel hydrogels based on methacrylate-modified agarose. Polym Int **51**(10), 867–870 (2002)
71. A. Tripathi, A. Kumar, Multi-featured macroporous agarose-alginate cryogel: Synthesis and characterization for bioengineering applications. Macromol. Biosci. **11**(1), 22–35 (2011)
72. H. Lin, A.W. Cheng, P.G. Alexander, A.M. Beck, R.S. Tuan, Cartilage Tissue engineering application of injectable gelatin hydrogel with in situ visible-light-activated gelation capability in both air and aqueous solution. Tissue Eng Part A **20**(17–18), 2402–2411 (2014)
73. X. Zhao, Q. Lang, L. Yildirimer, Z.Y. Lin, W. Cui, N. Annabi, et al., Photocrosslinkable gelatin hydrogel for epidermal tissue engineering. Adv. Healthc. Mater. **5**(1), 108–118 (2016)
74. A. Skardal, J. Zhang, L. McCoard, X. Xu, S. Oottamasathien, G.D. Prestwich, Photocrosslinkable hyaluronan-gelatin hydrogels for two-step bioprinting. Tissue Eng. Part A **16**(8), 2675–2685 (2010)
75. T. Billiet, B. Van Gasse, E. Gevaert, M. Cornelissen, J.C. Martins, P. Dubruel, Quantitative contrasts in the photopolymerization of acrylamide and methacrylamide-functionalized gelatin hydrogel building blocks. Macromol. Biosci. **13**(11), 1531–1545 (2013)
76. X. Li, S. Chen, J. Li, X. Wang, J. Zhang, N. Kawazoe, et al., 3D culture of chondrocytes in gelatin hydrogels with different stiffness. Polymers **8**(8), 269 (2016)
77. Y. Zhou, K. Liang, S. Zhao, C. Zhang, J. Li, H. Yang, et al., Photopolymerized maleilated chitosan/methacrylated silk fibroin micro/nanocomposite hydrogels as potential scaffolds for cartilage tissue engineering. Int. J. Biol. Macromol. **108**, 383–390 (2018)
78. I.S. Cho, M.O. Cho, Z. Li, M. Nurunnabi, S.Y. Park, S.-W. Kang, et al., Synthesis and characterization of a new photo-crosslinkable glycol chitosan thermogel for biomedical applications. Carbohydr. Polym. **144**, 59–67 (2016)
79. J.M. Jukes, L.J. van der Aa, C. Hiemstra, V. Tv, P.J. Dijkstra, Z. Zhong, et al., A newly developed chemically crosslinked dextran–poly(ethylene glycol) hydrogel for cartilage tissue engineering. Tissue Eng Part A **16**(2), 565–573 (2010)
80. K. Szafulera, R.A. Wach, A.K. Olejnik, J.M. Rosiak, P. Ulański, Radiation synthesis of biocompatible hydrogels of dextran methacrylate. Radiat. Phys. Chem. **142**, 115–120 (2018)
81. G. Chen, N. Kawazoe, Y. Ito, Photo-crosslinkable hydrogels for tissue engineering applications, in *Photochemistry for Biomedical Applications: From Device Fabrication to Diagnosis and Therapy*, ed. by Y. Ito, (Springer, Singapore, 2018), pp. 277–300
82. J.T. Oliveira, L. Martins, R. Picciochi, P.B. Malafaya, R.A. Sousa, N.M. Neves, et al., Gellan gum: A new biomaterial for cartilage tissue engineering applications. J. Biomed. Mater. Res. A **93**(3), 852–863 (2010)
83. M.A. Omobono, X. Zhao, M.A. Furlong, C.-H. Kwon, T.J. Gill, M.A. Randolph, et al., Enhancing the stiffness of collagen hydrogels for delivery of encapsulated chondrocytes to articular lesions for cartilage regeneration. J. Biomed. Mater. Res. A **103**(4), 1332–1338 (2015)
84. K. Yang, J. Sun, D. Wei, L. Yuan, J. Yang, L. Guo, et al., Photo-crosslinked mono-component type II collagen hydrogel as a matrix to induce chondrogenic differentiation of bone marrow mesenchymal stem cells. J. Mater. Chem. B **5**(44), 8707–8718 (2017)
85. K. Markstedt, A. Mantas, I. Tournier, H. Martinez Avila, D. Hagg, P. Gatenholm, 3D bioprinting human chondrocytes with nanocellulose-alginate bioink for cartilage tissue engineering applications. Biomacromolecules **16**(5), 1489–1496 (2015)

86. R.L. Mauck, M.A. Soltz, C.C. Wang, D.D. Wong, P.H. Chao, W.B. Valhmu, et al., Functional tissue engineering of articular cartilage through dynamic loading of chondrocyte-seeded agarose gels. J. Biomech. Eng. **122**(3), 252–260 (2000)
87. M.S. Ponticiello, R.M. Schinagl, S. Kadiyala, F.P. Barry, Gelatin-based resorbable sponge as a carrier matrix for human mesenchymal stem cells in cartilage regeneration therapy. J. Biomed. Mater. Res. **52**(2), 246–255 (2000)
88. S. Ibusuki, A. Papadopoulos, M.P. Ranka, G.J. Halbesma, M.A. Randolph, R.W. Redmond, et al., Engineering cartilage in a photochemically crosslinked collagen gel. J. Knee Surg. **22**(1), 72–81 (2009)
89. H. Tan, C.R. Chu, K.A. Payne, K.G. Marra, Injectable in situ forming biodegradable chitosan-hyaluronic acid based hydrogels for cartilage tissue engineering. Biomaterials **30**(13), 2499–2506 (2009)
90. J.K. Suh, H.W. Matthew, Application of chitosan-based polysaccharide biomaterials in cartilage tissue engineering: A review. Biomaterials **21**(24), 2589–2598 (2000)
91. W. Sun, B. Xue, Y. Li, M. Qin, J. Wu, K. Lu, et al., Polymer-supramolecular polymer double-network hydrogel. Adv Funct Mater **26**(48), 9044–9052 (2016)
92. Y.J. Seol, J.Y. Park, W. Jeong, T.H. Kim, S.Y. Kim, D.W. Cho, Development of hybrid scaffolds using ceramic and hydrogel for articular cartilage tissue regeneration. J. Biomed. Mater. Res. A **103**(4), 1404–1413 (2015)
93. B.R. Mintz, J.A. Cooper Jr., Hybrid hyaluronic acid hydrogel/poly(varepsilon-caprolactone) scaffold provides mechanically favorable platform for cartilage tissue engineering studies. J. Biomed. Mater. Res. A **102**(9), 2918–2926 (2014)
94. F.A. Formica, E. Ozturk, S.C. Hess, W.J. Stark, K. Maniura-Weber, M. Rottmar, et al., A bioinspired ultraporous nanofiber-hydrogel mimic of the cartilage extracellular matrix. Adv. Healthc. Mater. **5**(24), 3129–3138 (2016)
95. D. Chimene, K.K. Lennox, R.R. Kaunas, A.K. Gaharwar, Advanced bioinks for 3D printing: a materials science perspective. Ann Biomed Eng **44**(6), 2090–2102 (2016)
96. N. Naseri, B. Deepa, A.P. Mathew, K. Oksman, L. Girandon, Nanocellulose-based interpenetrating polymer network (IPN) hydrogels for cartilage applications. Biomacromolecules **17**(11), 3714–3723 (2016)
97. X. Zhang, G.J. Kim, M.G. Kang, J.K. Lee, J.W. Seo, J.T. Do, et al., Marine biomaterial-based bioinks for generating 3D printed tissue constructs. Mar Drugs **16**(12), 484 (2018)
98. D. Chimene, C.W. Peak, J.L. Gentry, J.K. Carrow, L.M. Cross, E. Mondragon, et al., Nanoengineered ionic–covalent entanglement (NICE) bioinks for 3D bioprinting. ACS Appl. Mater. Interfaces **10**(12), 9957–9968 (2018)
99. T. Lorson, S. Jaksch, M.M. Lubtow, T. Jungst, J. Groll, T. Luhmann, et al., A thermogelling supramolecular hydrogel with sponge-like morphology as a cytocompatible bioink. Biomacromolecules **18**(7), 2161–2171 (2017)
100. C.D. O'Connell, C. Di Bella, F. Thompson, C. Augustine, S. Beirne, R. Cornock, et al., Development of the biopen: A handheld device for surgical printing of adipose stem cells at a chondral wound site. Biofabrication **8**(1), 015019 (2016)
101. C. Onofrillo, S. Duchi, C.D. O'Connell, R. Blanchard, A.J. O'Connor, M. Scott, et al., Biofabrication of human articular cartilage: A path towards the development of a clinical treatment. Biofabrication **10**(4), 045006 (2018)
102. S.J. Bidarra, C.C. Barrias, P.L. Granja, Injectable alginate hydrogels for cell delivery in tissue engineering. Acta Biomater. **10**(4), 1646–1662 (2014)
103. K.S. Lim, R. Levato, P.F. Costa, M.D. Castilho, C.R. Alcala-Orozco, K.M.A. van Dorenmalen, et al., Bio-resin for high resolution lithography-based biofabrication of complex cell-laden constructs. Biofabrication **10**(3), 034101 (2018)
104. X. Pan, M.A. Tasdelen, J. Laun, T. Junkers, Y. Yagci, K. Matyjaszewski, Photomediated controlled radical polymerization. Prog. Polym. Sci. **62**, 73–125 (2016)
105. W.E. Hennink, C.F. van Nostrum, Novel crosslinking methods to design hydrogels. Adv. Drug Deliv. Rev. **54**(1), 13–36 (2002)

106. R.P. Rastogi, Richa, A. Kumar, M.B. Tyagi, R.P. Sinha, Molecular mechanisms of ultraviolet radiation-induced DNA damage and repair. J Nucleic Acids **2010**, 592980 (2010)
107. J. Cadet, E. Sage, T. Douki, Ultraviolet radiation-mediated damage to cellular DNA. Mutat Res **571**(1), 3–17 (2005)
108. J.-L. Ravanat, T. Douki, J. Cadet, Direct and indirect effects of UV radiation on DNA and its components. J. Photochem. Photobiol. B Biol. **63**(1), 88–102 (2001)
109. J.P.M. Wood, G. Lascaratos, A.J. Bron, N.N. Osborne, The influence of visible light exposure on cultured RGC-5 cells. Mol. Vis. **14**, 334–344 (2007)
110. E. Maverakis, Y. Miyamura, M.P. Bowen, G. Correa, Y. Ono, H. Goodarzi, Light, including ultraviolet. J. Autoimmun. **34**(3), J247–JJ57 (2010)
111. N. Eslahi, M. Abdorahim, A. Simchi, Smart polymeric hydrogels for cartilage tissue engineering: A review on the chemistry and biological functions. Biomacromolecules **17**(11), 3441–3463 (2016)
112. N.D. Tsihlis, J. Murar, M.R. Kapadia, S.S. Ahanchi, C.S. Oustwani, J.E. Saavedra, et al., Isopropylamine NONOate (IPA/NO) moderates neointimal hyperplasia following vascular injury. J. Vasc. Surg. **51**(5), 1248–1259 (2010)
113. J.J. Roberts, S.J. Bryant, Comparison of photopolymerizable thiol-ene PEG and acrylate-based PEG hydrogels for cartilage development. Biomaterials **34**(38), 9969–9979 (2013)
114. F. Tanaka, S.F. Edwards, Viscoelastic properties of physically crosslinked networks: Part 1. Non-linear stationary viscoelasticity. J. Non-Newtonian Fluid Mech. **43**(2), 247–271 (1992)
115. L. Zhao, H. Mitomo, F. Yosh, Synthesis of pH-sensitive and biodegradable CM-cellulose/chitosan polyampholytic hydrogels with electron beam irradiation. J Bioact Compat Pol **23**(4), 319–333 (2008)
116. T. Miyazaki, Y. Takeda, S. Akane, T. Itou, A. Hoshiko, K.J.P. En, Role of boric acid for a poly (vinyl alcohol) film as a cross-linking agent: Melting behaviors of the films with boric acid. Polymer **51**(23), 5539–5549 (2010)
117. F. Ullah, M.B.H. Othman, F. Javed, Z. Ahmad, H.M. Akil, Classification, processing and application of hydrogels: A review. Mater Sci Eng C Mater Biol Appl **57**, 414–433 (2015)
118. R.F. Pereira, P.J. Bártolo, 3D photo-fabrication for tissue engineerying and drug delivery. Engineering **1**(1), 090–112 (2015)
119. F. Jivan, N. Fabela, Z. Davis, D.L. Alge, Orthogonal click reactions enable the synthesis of ECM-mimetic PEG hydrogels without multi-arm precursors. J Mater Chem B **6**(30), 4929–4936 (2018)
120. H. Shih, C.C. Lin, Cross-linking and degradation of step-growth hydrogels formed by thiol–ene photoclick chemistry. Biomacromolecules **13**(7), 2003–2012 (2012)
121. I.-M. Yu, F.M. Hughson, Tethering factors as organizers of intracellular vesicular traffic. Annu Rev Cell Dev Biol **26**, 137–156 (2010)
122. L.-S. Wang, C. Du, W.S. Toh, A.C. Wan, S.J. Gao, M. Kurisawa, Modulation of chondrocyte functions and stiffness-dependent cartilage repair using an injectable enzymatically cross-linked hydrogel with tunable mechanical properties. Biomaterials **35**(7), 2207–2217 (2014)
123. L. Bian, C. Hou, E. Tous, R. Rai, R.L. Mauck, J.A. Burdick, The influence of hyaluronic acid hydrogel crosslinking density and macromolecular diffusivity on human MSC chondrogenesis and hypertrophy. Biomaterials **34**(2), 413–421 (2013)
124. C. Chung, J. Mesa, M.A. Randolph, M. Yaremchuk, J.A. Burdick, Influence of gel properties on neocartilage formation by auricular chondrocytes photoencapsulated in hyaluronic acid networks. J. Biomed. Mater. Res. A **77**(3), 518–525 (2006)
125. A.J. Neumann, T. Quinn, S.J. Bryant, Nondestructive evaluation of a new hydrolytically degradable and photo-clickable PEG hydrogel for cartilage tissue engineering. Acta Biomater. **39**, 1–11 (2016)
126. G.C. Ingavle, S.H. Gehrke, M.S. Detamore, The bioactivity of agarose-PEGDA interpenetrating network hydrogels with covalently immobilized RGD peptides and physically entrapped aggrecan. Biomaterials **35**(11), 3558–3570 (2014)

127. S.C. Skaalure, S.O. Dimson, A.M. Pennington, S.J. Bryant, Semi-interpenetrating networks of hyaluronic acid in degradable PEG hydrogels for cartilage tissue engineering. Acta Biomater. **10**(8), 3409–3420 (2014)
128. C.B. Rodell, N.N. Dusaj, C.B. Highley, J.A. Burdick, Injectable and cytocompatible tough double-network hydrogels through tandem supramolecular and covalent crosslinking. Adv Mater **28**(38), 8419–8424 (2016)
129. H. Jung, J.S. Park, J. Yeom, N. Selvapalam, K.M. Park, K. Oh, et al., 3D tissue engineered supramolecular hydrogels for controlled chondrogenesis of human mesenchymal stem cells. Biomacromolecules **15**(3), 707–714 (2014)
130. K. Wei, M. Zhu, Y. Sun, J. Xu, Q. Feng, S. Lin, et al., Robust biopolymeric supramolecular "host−guest macromer" hydrogels reinforced by in situ formed multivalent nanoclusters for cartilage regeneration. Macromolecules **49**(3), 866–875 (2016)
131. Y. Guo, T. Yuan, Z. Xiao, P. Tang, Y. Xiao, Y. Fan, et al., Hydrogels of collagen/chondroitin sulfate/hyaluronan interpenetrating polymer network for cartilage tissue engineering. J. Mater. Sci. Mater. Med. **23**(9), 2267–2279 (2012)
132. T. Wang, J.H. Lai, F. Yang, Effects of hydrogel stiffness and extracellular compositions on modulating cartilage regeneration by mixed populations of stem cells and chondrocytes in vivo. Tissue Eng. Part A **22**(23–24), 1348–1356 (2016)
133. S.J. Bryant, K.S. Anseth, Hydrogel properties influence ECM production by chondrocytes photoencapsulated in poly(ethylene glycol) hydrogels. J. Biomed. Mater. Res. **59**(1), 63–72 (2002)
134. O. Chaudhuri, L. Gu, D. Klumpers, M. Darnell, S.A. Bencherif, J.C. Weaver, et al., Hydrogels with tunable stress relaxation regulate stem cell fate and activity. Nat. Mater. **15**(3), 326–334 (2016)
135. M. Darnell, S. Young, L. Gu, N. Shah, E. Lippens, J. Weaver, et al., Substrate stress-relaxation regulates scaffold remodeling and bone formation in vivo. Adv Healthcare Mater **6**(1) 1601185 (2017)
136. H.P. Lee, L. Gu, D.J. Mooney, M.E. Levenston, O. Chaudhuri, Mechanical confinement regulates cartilage matrix formation by chondrocytes. Nat. Mater. **16**(12), 1243–1251 (2017)
137. B.M. Richardson, D.G. Wilcox, M.A. Randolph, K.S. Anseth, Hydrazone covalent adaptable networks modulate extracellular matrix deposition for cartilage tissue engineering. Acta Biomater. **83**, 71–82 (2019)
138. S.C. Neves, R.F. Pereira, M. Araújo, C.C. Barrias, Bioengineered peptide-functionalized hydrogels for tissue regeneration and repair, in *Peptides and Proteins as Biomaterials for Tissue Regeneration and Repair*, (Woodhead Publishing, Sawston, Cambridge, UK, 2018), pp. 101–125
139. R.F. Pereira, A. Sousa, C.C. Barrias, P.J. Bártolo, P.L. Granja, A single-component hydrogel bioink for bioprinting of bioengineered 3D constructs for dermal tissue engineering. Mater Horiz **5**(6), 1100–1111 (2018)
140. H.J. Lee, C. Yu, T. Chansakul, N.S. Hwang, S. Varghese, S.M. Yu, et al., Enhanced chondrogenesis of mesenchymal stem cells in collagen mimetic peptide-mediated microenvironment. Tissue Eng. Part A **14**(11), 1843–1851 (2008)
141. C.N. Salinas, K.S. Anseth, Decorin moieties tethered into PEG networks induce chondrogenesis of human mesenchymal stem cells. J. Biomed. Mater. Res. A **90**(2), 456–464 (2009)
142. J.T. Connelly, A.J. Garcia, M.E. Levenston, Inhibition of in vitro chondrogenesis in RGD-modified three-dimensional alginate gels. Biomaterials **28**(6), 1071–1083 (2007)
143. I. Villanueva, C.A. Weigel, S.J. Bryant, Cell-matrix interactions and dynamic mechanical loading influence chondrocyte gene expression and bioactivity in PEG-RGD hydrogels. Acta Biomater. **5**(8), 2832–2846 (2009)
144. R.F. Pereira, C.C. Barrias, P.J. Bartolo, P.L. Granja, Cell-instructive pectin hydrogels crosslinked via thiol-norbornene photo-click chemistry for skin tissue engineering. Acta Biomater. **66**, 282–293 (2018)

145. P.A. Parmar, L.W. Chow, J.P. St-Pierre, C.M. Horejs, Y.Y. Peng, J.A. Werkmeister, et al., Collagen-mimetic peptide-modifiable hydrogels for articular cartilage regeneration. Biomaterials **54**, 213–225 (2015)
146. C. Bonnans, J. Chou, Z. Werb, Remodelling the extracellular matrix in development and disease. Nat. Rev. Mol. Cell Biol. **15**(12), 786–801 (2014)
147. G.S. Schultz, A. Wysocki, Interactions between extracellular matrix and growth factors in wound healing. Wound repair and regeneration: Official publication of the Wound Healing Society [and] the European Tissue Repair. Society **17**(2), 153–162 (2009)
148. B.V. Sridhar, N.R. Doyle, M.A. Randolph, K.S. Anseth, Covalently tethered TGF-beta1 with encapsulated chondrocytes in a PEG hydrogel system enhances extracellular matrix production. J. Biomed. Mater. Res. A **102**(12), 4464–4472 (2014)
149. Q. Feng, S. Lin, K. Zhang, C. Dong, T. Wu, H. Huang, et al., Sulfated hyaluronic acid hydrogels with retarded degradation and enhanced growth factor retention promote hMSC chondrogenesis and articular cartilage integrity with reduced hypertrophy. Acta Biomater. **53**, 329–342 (2017)
150. L. Bian, D.Y. Zhai, E. Tous, R. Rai, R.L. Mauck, J.A. Burdick, Enhanced MSC chondrogenesis following delivery of TGF-beta3 from alginate microspheres within hyaluronic acid hydrogels in vitro and in vivo. Biomaterials **32**(27), 6425–6434 (2011)
151. B.V. Sridhar, J.L. Brock, J.S. Silver, J.L. Leight, M.A. Randolph, K.S. Anseth, Development of a cellularly degradable PEG hydrogel to promote articular cartilage extracellular matrix deposition. Adv. Healthc. Mater. **4**(5), 702–713 (2015)

Chapter 8
Exoskeletons for Lower Limb Applications: A Review

Mohammed S. Alqahtani, Glen Cooper, Carl Diver, and Paulo Jorge Bártolo

8.1 Introduction

Exoskeletons can be defined as wearable or robotic devices that assist people in performing their daily life movements, thus boosting the user's performance. They are used for regaining mobility, thereby allowing people to walk, stand and sit [13]. Exoskeletons, consisting of sensors, actuators and control elements, are utilised in various applications requiring carrying heavy loads or for rehabilitation and assistance of paralysed patients [32, 34].

Different classifications of exoskeletons have been proposed as shown in Fig. 8.1 [1, 12, 30, 34]. One classification is based on the part of the body the exoskeleton supports. This classifies exoskeletons as an upper limb, lower limb or full body. Exoskeletons can also be classified as active, passive or quasi-passive. Active exoskeletons require an energy source to actuate sensors and actuators, whereas passive devices do not require any energy source as they are only formed by mechanical elements such as linkage, springs and dampers. The quasi-passive devices lie between these two types [53]. Depending on the application, exoskeletons can be classified as devices for gait rehabilitation, human locomotion assistance and human strength augmentation. Finally, depending on the type of actuators, exoskeletons can be classified as electric, pneumatic or hydraulic actuators.

M. S. Alqahtani · G. Cooper · P. J. Bártolo (✉)
School of Mechanical, Aerospace and Civil Engineering, The University of Manchester, Manchester, UK
e-mail: mohammed.alqahtani-4@postgrad.manchester.ac.uk; glen.cooper@manchester.ac.uk; paulojorge.dasilvabartolo@manchester.ac.uk

C. Diver
Department of Engineering, Manchester Metropolitan University, Manchester, UK
e-mail: c.diver@mmu.ac.uk

P. J. Bártolo, B. Bidanda (eds.), *Bio-Materials and Prototyping Applications in Medicine*, https://doi.org/10.1007/978-3-030-35876-1_8

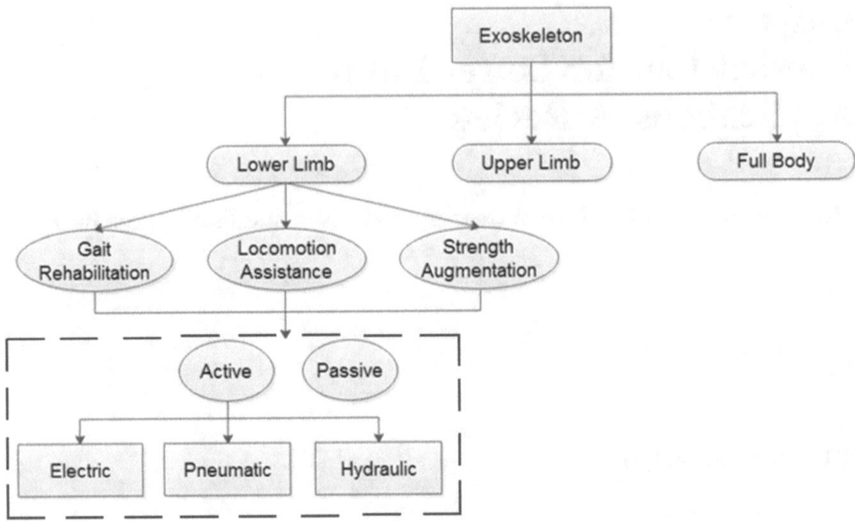

Fig. 8.1 Classification of exoskeletons with particular emphasis on lower limb systems

This is a very active research domain. According to a research through PubMed (July 2018), 140 papers were published under the topic lower limb exoskeleton design since 2003 (44% published in the last 2 years). The majority of the papers published (32 papers) focuses on control aspects including safety control, with particular emphasis on mechanical design aspects (12% of the papers). Human interface aspects are covered by 5% of the papers, around 7% covered biomechanical considerations (e.g. kinematic, compatibility, the range of motion) and 14% covered topics such as customisation, low-weight aspects, pressure reduction, actuators and energy expenditure.

This chapter focuses on the current state of the art of lower limb exoskeletons. It starts by presenting a few biomechanics concepts regarding the lower limb part of the human body, which allows understanding of the complex nature of the movements and forces that must be considered to design a proper exoskeleton. Then, the different classes of exoskeleton are described in a detailed way. Research challenges and future perspectives are also presented.

8.2 Lower Limb Biomechanics and Locomotion

Understanding human locomotion and the anatomy of the lower limb are essential in the design of exoskeletons [22]. The human lower extremity consists of three main joints: the hip, the knee and the ankle joints [54], shown in Fig. 8.2. The human lower limb can be simplified to seven degrees of freedom (DOFs) (three at the hip, one at the knee [55] and three at the ankle [79]) in different planes of the body [22]. Degrees of freedom and planes for lower limb motion are shown in Table 8.1.

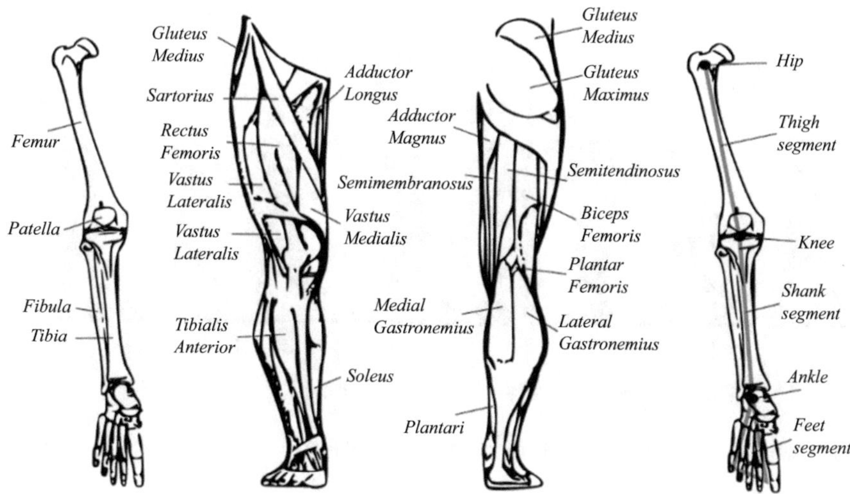

Fig. 8.2 Lower limb anatomy [56]

Table 8.1 The allowed motions at lower limb joints in different planes of the body

Joint	Plane	Motion
Hip	Sagittal	Flexion
		Extension
	Coronal	Adduction
		Abduction
	Transverse	Internal rotation
		External rotation
Knee	Sagittal	Extension
		Flexion
Ankle	Sagittal	Dorsiflexion
		Plantarflexion
	Transverse	Adduction
		Abduction
	Coronal	Inversion
		Eversion

Movement is a complex neural and biomechanical process, determined by the interaction between the central nervous system, peripheral nervous system and the musculoskeletal system [67]. Human gait is an essential part of daily living. It consists of two phases (stance phase and swing phase) and is enabled by joint movement in the lower limb. The designers of exoskeletons need to understand these joints kinematics and the gait cycle to develop effective devices. The stance phase is divided into four intervals, namely, the loading response (LRP), mid-stance (MST), terminal stance (TST) and pre-swing (PSW) [56]. The swing phase comprises three key periods, namely, initial swing (ISW), mid-swing (MSW) and terminal swing (TSW). The human gait cycle is illustrated in Fig. 8.3.

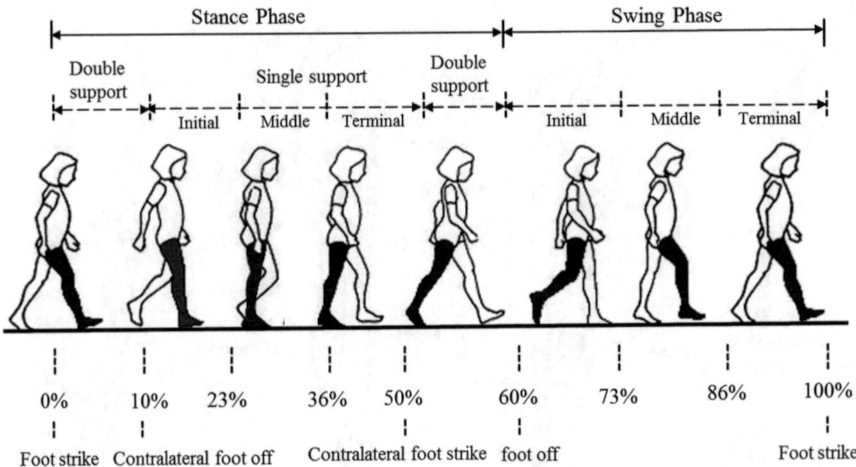

Fig. 8.3 Human gait cycle

The hip joint is located between the head of the femur and the acetabulum of the pelvis [56]. It is a ball and socket stable and strong joint, with three DOFs, surrounded by strong muscles able to perform flexion/extension, abduction/adduction, internal (medial)/external (lateral) rotations. Flexion/extension is considered to be the main DOF used in the locomotive activity [11]. The knee joint formed by the femur, tibia, fibula and patella has a number of functions during walking [56]. It comprises three articulations (tibiofemoral joint, patellofemoral joint and tibiofibular joint) and can be simplified to one DOF, allowing flexion/extension in the sagittal plane [31, 55]. The ankle joint, formed by the connection of tibia, fibula and talus, plays an important role in the equilibrium of the lower limb system [56]. This joint has three DOFs: plantarflexion/dorsiflexion, adduction/abduction and inversion/eversion [31, 79].

8.3 Classes of Lower Limb Exoskeleton

8.3.1 Gait Rehabilitation, Human Locomotion Assistance and Human Strength Augmentation

8.3.1.1 Gait Rehabilitation

Rehabilitation is an important treatment to improve the recovery of the lower limb motor functions of patients suffering from neurological disorder such as stroke, Parkinson's disease, traumatic brain injury, spinal cord injuries, muscular dystrophy, spinal cord atrophy or cerebral palsy, enabling them to walk independently [29, 43, 60, 64]. All of these disorders result in muscular weakness which is the main reason for the development of rehabilitation exoskeletons [29]. In these cases, manual rehabilitation process is a complicated task requiring significant efforts from both the therapist and patient. Moreover, manual rehabilitation cannot provide intensive training, and the training time is limited to the therapist availability. Due to these difficulties, robotic rehabilitation devices are increasingly being used as an ideal solution for repetitive tasks, allowing the therapist to focus on other tasks such as analysing the gait performance of the patient [12, 15].

Gait rehabilitation exoskeletons can be classified into two main groups: treadmill-based exoskeletons and overground exoskeletons [43]. Treadmill-based exoskeletons are immobile robots that provide gait rehabilitation in a fixed and confined area. By contrast, overground rehabilitation robots are mobile and designed to allow patients walking over the ground in unrestricted areas. This makes the patient more independent when performing gait training. Moreover, overground exoskeletons allow patients to regain natural gait [15].

The treadmill is a rehabilitation technique that is used to improve mobility functions of patients and to improve their ability to walk after brain injury [21]. It is used to train patients with cardiopulmonary diseases and also for the rehabilitation of patients with orthopaedic and neurological diseases [10]. This exoskeleton consists of two-powered leg orthoses, body weight support system and treadmill [15]. They are stationary exoskeletons that have a fixed structure and a mobile ground platform [9]. Examples of commercially available systems include the Lokomat (Hocoma, Inc., Switzerland), LokoHelp (LokoHelp Group, Germany) and ReoAmbulator (Motorika Ltd., USA) [21, 23, 27, 71]. Among them, the Lokomat is the most commonly used device (Fig. 8.4a, b). On the one hand, it combines a physical exoskeleton with a virtual reality environment of audio and visual biofeedback and uses a DC motor with helical gears to precisely control the trajectory of the hip and knee joints [15, 23]. On the other hand, the overground exoskeleton is a mobile robotic base, consisting of robots that follow the motions of the patient's walking on overground. Rather than making the patient follow predetermined movements, this system allows patients to move under their control [21]. A number of overground gait trainers have already been commercialised such as WalkTrainer (Swortec SA, Switzerland) ReWalk (ARGO Medical), eLEGS and Indego [15, 21, 60].

The WalkTrainer (Fig. 8.4c, d), which provides overground walking with control of the pelvic motion, is composed of five main components: frame, body weight support, two leg orthoses, pelvic orthosis and electro-stimulator [6]. The main function of the frame is to follow the patient during the walking exercise. The body weight support system is used to prevent the patient from falling, the leg orthosis measures the positions of the hip, knee and ankle joints, monitoring the interaction forces between the leg orthosis and the patient, and the pelvic orthosis assists the patient during walking. Finally, the primary function of the electro-stimulator is to stimulate some muscles (e.g. gluteus maximus, biceps femoris, rectus femoris, vastus lateralis and medialis, tibialis anterior and gastrocnemius) in the patient [6, 15].

The Active Knee Rehabilitation Orthotic System (ANdROS) developed by Unluhisarcikli et al. [63] is another example of a portable gait rehabilitation. This exoskeleton (Fig. 8.5) allows for motor control by applying a corrective torque around the knee joint using an impedance controller. The device also contains two ankle-foot orthoses rigidly attached to the main frame.

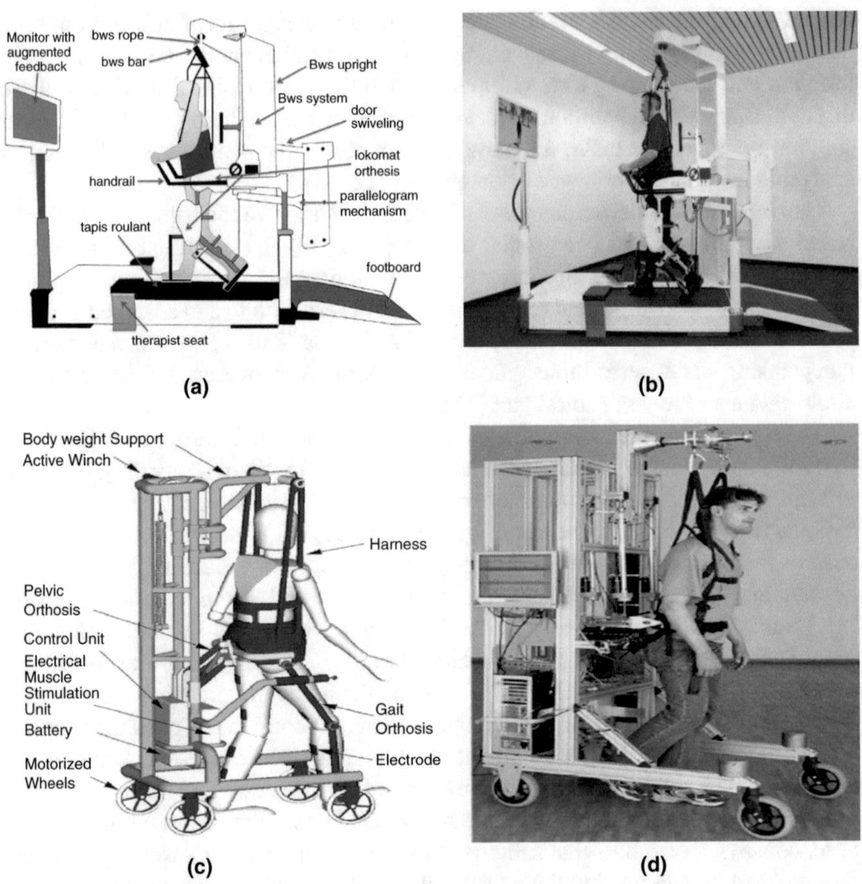

Fig. 8.4 (**a**) Schematic of Lokomat robotic system with its components [9]. (**b**) Lokomat system with the patient [21]. (**c**) The WalkTrainer robotic device with its components. (**d**) A patient testing the WalkTrainer device [6]

The ReWalk exoskeleton (Fig. 8.6) consists of a brace support suit, battery, sensors to measure the tilt angle of the upper body, joint angles and the contact with the ground and a computerised system located in the backpack. It is powered by a DC motor at the hip joint and the knee joint, while the ankle joint is un-actuated. The controller and the battery are attached to the back of the exoskeleton [43]. In addition to the system, patients should use crutches to maintain balance control [15, 60].

The eLEGS exoskeleton built in 2010 and renamed as Ekso in 2011 is commercialised by Ekso Bionics (USA). It is an exoskeleton designed for patients with hemiplegia due to stroke or spinal cord injury, presenting a total of six DOFs (3 DOFs per leg) [28] (Fig. 8.7a). The hip and the knee joints are driven by electric actuators, while the ankle joints are passive [12, 59]. The device also has three straps on each leg to support legs and a backpack that contains the battery and the controller. Crutches should also be used to support the user and to control the exoskeleton.

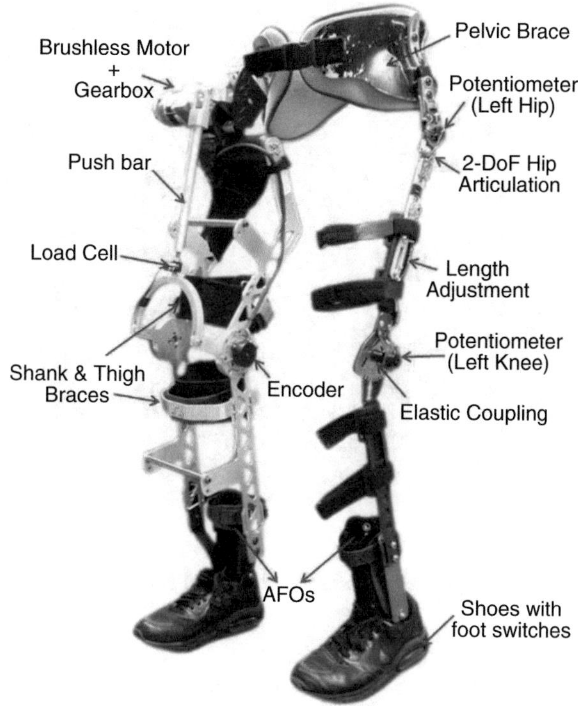

Fig. 8.5 The ANdROS lower limb exoskeleton [64]

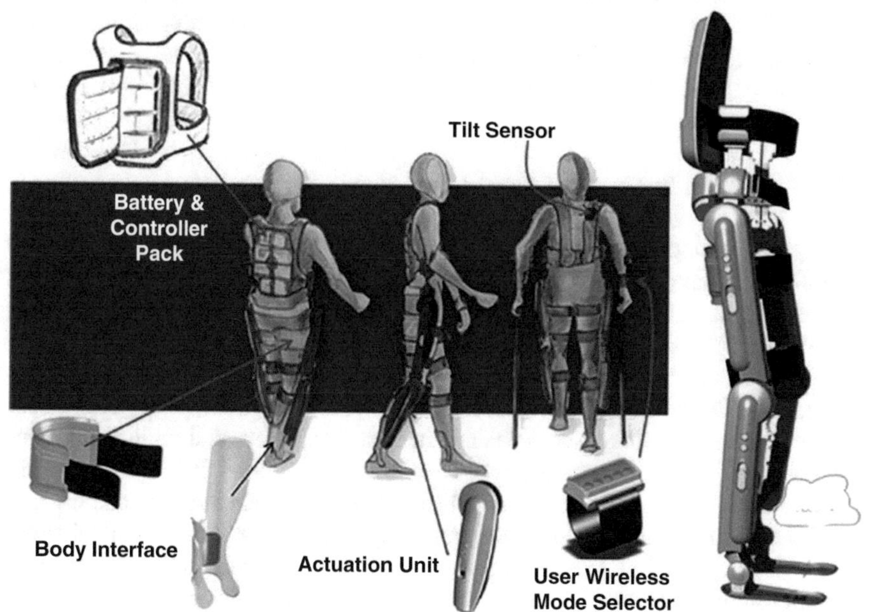

Fig. 8.6 The configuration of the ReWalk exoskeleton [25]

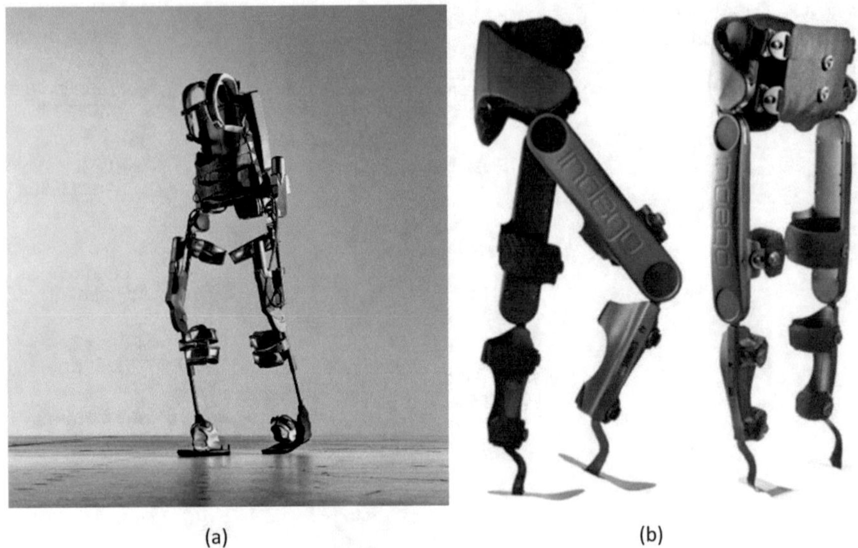

Fig. 8.7 (**a**) The Ekso exoskeleton [59]. (**b**) The Indego exoskeleton [46]

The Indego system (Fig. 8.7b) consists of three main components: the hip brace which contains the battery and the controller, two thigh frames and two shank frames. These parts can be easily split and then assembled again [59]. The hip and knee joints are electronically actuated by DC brushless motor, with passive ankle joints [28]. The Indego system allows a number of actions including sitting, standing, walking, sit to stand, stand to sit, walk to stand and stand to walk. It also has brakes at the knee joints to prevent knee buckling in the case of power failure [18].

8.3.1.2 Human Locomotion Assistance

This type of exoskeletons enables to restore mobility in patients with paraplegia as a result of spinal cord injuries [53]. These exoskeletons provide a wide range of movements that allow users to perform daily life motions such as walking, standing and sitting [70]. Human locomotion assistance exoskeletons can help paralysed patients regain their mobility, thus improving their mental and physical health [13]. An example is the walking power assist leg (WPAL) (Fig. 8.8a), designed to help people who suffer from muscle weakness in their lower limbs and for human power augmentation [2, 14]. The design consists of 12 DOFs in total, 3 DOFs at each hip joint (flexion/extension, abduction/adduction and internal/external rotations), 1 DOF at each knee joint (flexion/extension), 1 DOF at each ankle joint (dorsiflexion/plantarflexion) and 1 DOF at each metatarsophalangeal joint. The hip joints (flexion/extension) and knees (flexion/extension) are actuated by using a DC servomotor coupled with a harmonic reducer gear, whereas the other joints are free [2].

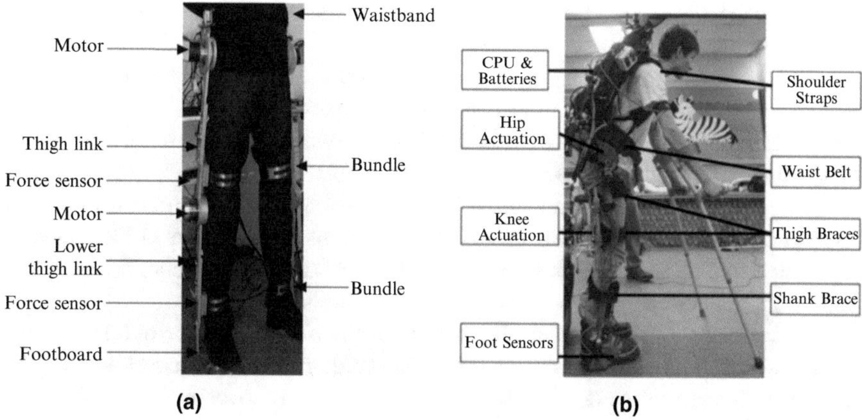

Fig. 8.8 (**a**) The WPAL exoskeleton [14]. (**b**) The medical exoskeleton with its component [62]

The Asian Institute of Technology Leg Exoskeleton-I (ALEX-I) was designed to support paraplegic persons. It consists of 12 DOFs, 3 DOFs at each hip, 1 DOF at each knee and 2 DOFs at each ankle. The exoskeleton is driven by 12 DC motors, at the hips (flexion/extension, abduction/adduction, internal/external rotations), at the knees (flexion/extension) and the ankles (dorsiflexion/plantarflexion, abduction/adduction) [3].

The lower limb orthosis proposed by Wu et al. [73] was designed to provide the wearer with full assistance when performing daily activities. The knee joint (flexion/extension) and ankle joint (dorsiflexion/plantarflexion) are driven by a pneumatic actuator, and the device aims to provide 100% torque to a person of 75 kg of weight. Strausser et al. [62] developed a medical exoskeleton which is shown in Fig. 8.8b for paraplegic mobility with 8 DOFs utilising hydraulic actuators at hip joint (flexion/extension) and knee joint (flexion/extension) to move patient's joints and utilising passive springs at the hip joint (internal/external rotations) and ankle joint (dorsiflexion/plantarflexion).

8.3.1.3 Human Strength Augmentation

Human strength augmentation exoskeletons intend to amplify the physical abilities of the users. They improve human endurance and strength during locomotion, allowing the users to carry heavy loads and walk for long distances. Moreover, they provide the wearer with strength to perform laborious tasks [12]. These exoskeletons are designed for several applications: material handling in harmful environment, assistive devices for disabled patients, industrial and military fields, disaster relief workers, carrying heavy payloads and rescuing of victims [12, 45]. Different designs have been proposed. Yamamoto et al. [74] developed the so-called power assisting suit, to support the work of a nurse in carrying a patient by his/her arm. The device consists of the shoulders, arms, waist and legs made of aluminium in order to create a lightweight structure. The arms, waist and legs are motorised using pneumatic rotary actuators.

The hydraulic lower extremity exoskeleton robot, developed by Kim et al. [39], was designed to enable soldiers to carry heavy loads on their back and to reduce muscle fatigue caused by these loads. It allows the wearer to carry a maximum load of 45 kg with a speed of 4 km/h. The design has a total of 12 DOFs. It has four active joints powered by hydraulic actuators (1 DOF at each hip (flexion/extension) and 1 DOF at each knee joint (flexion/extension)) and eight passive joints (1 DOF at each hip (internal/external rotations) and 3 DOFs for each ankle (dorsiflexion/plantarflexion, abduction/adduction, internal/external rotations)).

The Nanyang Technological University built a wearable lower extremity exoskeleton (NTU-LEE) based on an inner exoskeleton and outer exoskeleton. The inner exoskeleton uses encoders to measure the human movement [44], while the outer exoskeleton tracks the encoder signals using a proportional integral derivative (PID) controller during the load carrying process. The outer exoskeleton was designed with a total of 11 DOFs. Motors and encoders are used at hips (flexion/extension), knees (flexion/extension) and ankles (dorsiflexion/plantarflexion) with linear actuators at the trunk. Springs are also applied at the hips (abduction/adduction) and ankles (abduction/adduction) [2].

8.4 Active Versus Passive

The history of active devices dates back to the 1960s when the US military introduced several exoskeletons for military purposes. Most of these exoskeleton systems are active and use electric motors that require continuous power input [48]. As discussed by Yli-Peltola [76], most exoskeletons are powered by electricity or pressurised air, enabling movement through sensors and actuators [47].

Active exoskeletons are defined as devices that require actuators to apply forces on the legs allowing movement [5]. One of the most common is based on the muscles' electromyography (EMG) concept [1, 57]. The aim is to reduce muscle recruitment in the lower limb during locomotion which is usually measured by EMG signals [77]. The EMG signal is a biomedical signal that measures the electrical current generated during the contracting of muscles. This signal is controlled by the nervous system and is dependent on numerous anatomical and physiological properties of muscles (motor units, muscle fibres, specialised cells, extensibility, elasticity, contraction, relaxation) [57].

In order to acquire EMG data, electrodes need to be placed on the skin over the muscles [51]. EMG signals are used to evaluate the intended motion of the user. Additionally, they can be used to measure the level of interaction between the human and the exoskeleton and to control the actuators of the device [1, 49]. By analysing the muscles of a lower limb, the EMG data could be helpful for developing a lower limb exoskeleton [51]. EMG activity has been used in some devices such as the LOPES and the HAL [1, 36, 69]. The lower extremity-powered exoskeleton (LOPES) (Fig. 8.9a, b) was initially developed by Ekkelenkamp et al. [24] and further improved by Veneman et al. [69]. It has a total of 8 DOFs, two actuated pelvis segments and three actuated joints for each leg, two at the hip joint (flexion/extension, adduction/abduction) and one at the knee joint (flexion/extension). It uses EMG signals to measure muscle activity and to predict the user's motion [69]. The LOPES exoskeleton is actuated by using Bowden cable driven series elastic actuators and impedance controlled, allowing bidirectional mechanical interactions between the exoskeleton and the user [58].

The hybrid assistive limb (HAL) is a wearable device developed by Tsukuba University in Japan [29, 36]. The system consists of three main components, the skeleton and actuator, controller and sensors as shown in Fig. 8.9c. The frame, made of aluminium alloy and steel, is attached to the external part of the lower limb of the patient [36]. The HAL system has 4 DOFs actuated by using harmonic drive gear

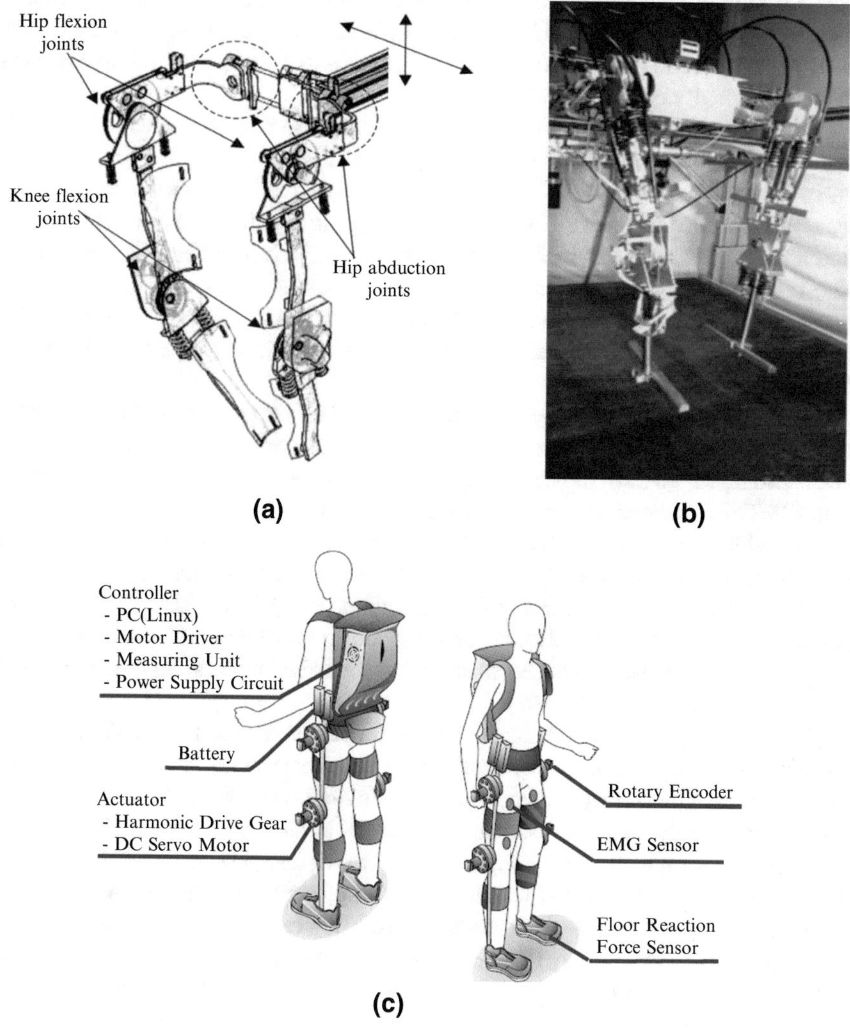

Fig. 8.9 (**a**) Design of the LOPES exoskeleton [68]. (**b**) LOPES exoskeleton prototype [15]. (**c**) The basic elements of HAL-3 exoskeleton [36]

and DC servo motor to generate torque at the hip and knee joints [2, 37]. It also uses EMG signal to predict the intended motion of the patient [36].

On the one hand, these active devices bring many merits to the user. They can offer an assistance to move the body of the patient suffering from spinal cord injuries or other diseases. Moreover, it can help the user to effectively, easily and quickly perform a certain movement [76]. As the majority of exoskeletons available in the market are active devices either electrically or pneumatically powered, this might lead to an increase in the weight of the exoskeleton resulting from the large weight of motors and power supplies. Consequently, active exoskeletons are not the better choice for training stroke patients [7].

On the other hand, passive devices heavily rely on the ability of the user to apply forces to move the leg [5]. Based on the requirements from the US army to have devices without recharging, the actuators and power sources were removed, reducing the weight of the device. They operate by transferring the load from the device and the backpack to the ground by a frame [76]. Passive devices use passive springs to provide gravity compensation [48].

An example of unpowered exoskeletons is the passive ankle exoskeleton developed by Collins et al. [17]. It is a lightweight exoskeleton designed to reduce the energy cost of human walking. The device is made of a carbon fibre frame, mechanical clutch, cable and spring (Fig. 8.10a). The spring is parallel to the Achilles tendon and attached to the human leg by the frame and a lever about the ankle joint. The function of the mechanical clutch which is parallel to the calf muscles is to engage the spring during heel strike and disengage when the foot is in the air, allowing for free motion. The allowed movements at the ankle joint are plantarflexion and dorsiflexion [17].

The XPED 2 (Fig. 8.10b) is a passive exoskeleton developed by Van Dijk et al. [66] to reduce joint torques during walking. It uses elastic elements known as artificial tendons, which have the ability to store and transfer energy between joints [66]. The XPED 2 is formed by a rigid frame connected to the human body at the pelvis, shank and foot; a backpack and a cable that extends from a lever at the pelvis to a leaf spring at the foot via a pulley at the knee. The primary function of the leaf spring is to provide the elasticity to the exoskeleton [65, 66]. The device has 6 DOFs per leg (flexion/extension, abduction/adduction, internal/external rotations at the hip joint; flexion/extension at the knee joint; plantarflexion/dorsiflexion, pronation/supination at the ankle joint) [65].

Quasi-passive devices lie between active and passive exoskeletons. They require a small power supply unit to operate electronic control systems, clutches or variable dampers [53]. The concept of this light and efficient type of exoskeletons seeks to use the passive dynamics of human walking [54]. The MoonWalker (Fig. 8.10c) developed by Krut et al. [41] is a good example of quasi-passive exoskeletons. It is a lower limb exoskeleton that has the ability to partially sustain the user's weight through the use of a passive force balancer. This device is controlled by an actuator that requires very low energy to work. This is only used to shift that force the same side as the leg in stance. The exoskeleton can be used for rehabilitation and also as an assistive device.

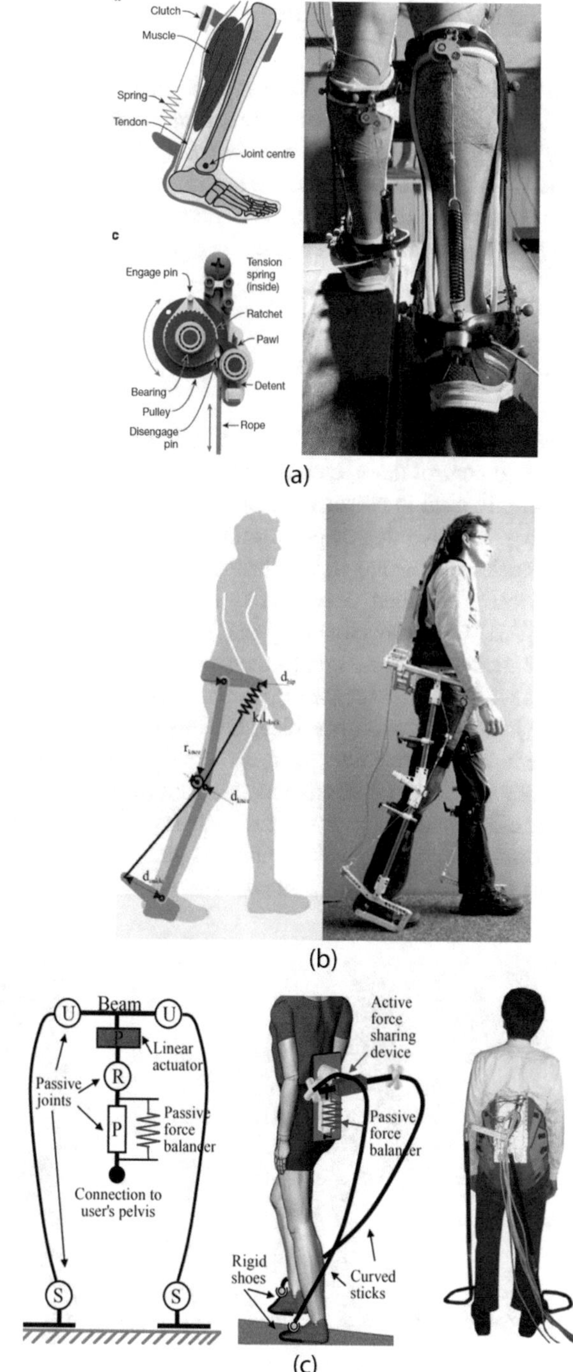

Fig. 8.10 (**a**) The components of the passive ankle exoskeleton [17]. (**b**) The XPED 2 exoskeleton [66]. (**c**) The MoonWalker lower limb exoskeleton [41]

8.5 Actuation Systems

The actuation system consists of three different types of actuators: electric, pneumatic and hydraulic. The main function of these actuators is to provide the necessary power to the exoskeleton, enabling it to perform a certain task [30]. The majority of the exoskeletons are actuated using electric actuators (65%), and 27% uses pneumatic actuators [30, 52]. A combination of multiple actuators (hybrid system) has been also explored [2].

8.5.1 Electric Actuators

In order to address the bulkiness of hydraulic and pneumatic actuators and the difficulty to control them, electric motors are extensively used, making them the most commonly used in exoskeleton systems [15, 30]. They are easily controllable, provide a good and quick response and present the lowest power-to-weight ratio. Major drawbacks are related to backlash and friction [8, 15, 30]. A good example of an electrically actuated exoskeleton is the Hanyang Exoskeleton Assistive Robot (HEXAR)-CR50 developed by Lim et al. [42]. This exoskeleton (Fig. 8.11a) aimed to enhance muscle strength of the user while transporting a load. It has a total of 14 DOFs with 3 DOFs for the hip joints, 1 DOF for the knee joints and 3 DOFs for the ankle joints per foot. A brushless DC electric motor and harmonic gear were applied at the hip and knee joints for flexion and extension movement, while the ankle and toe joints use a quasi-passive mechanism [42].

Another example is the Vanderbilt lower limb orthosis (Fig. 8.11b) which is a powered device designed to provide gait assistance to patients with spinal cord injury. The device was designed to assist the flexion and extension of the hip and

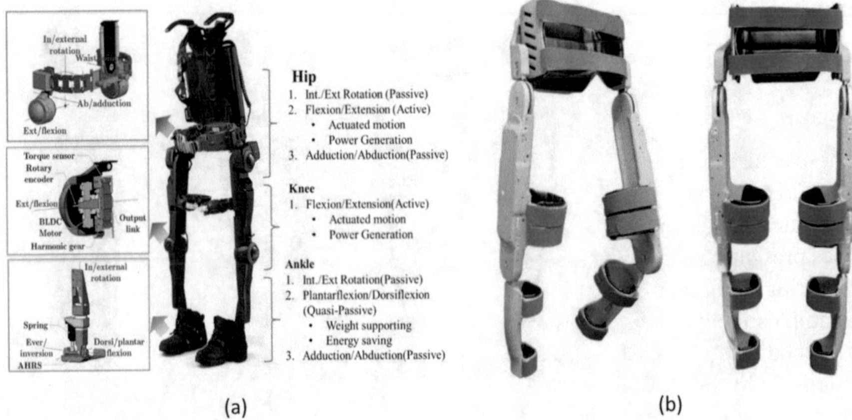

Fig. 8.11 (**a**) Structure of HEXAR-CR50 exoskeleton and joint modules [42]. (**b**) The structure of the Vanderbilt [26]

knee joints, which are actuated by brushless DC motor [75]. The knee motors are equipped with electrically locked brakes to keep knee joint locked during a power failure. The brake is locked during the stance phase and unlocked during the swing phase also from sit to stand and stand to sit. The device includes potentiometers in the hip and knee joints, accelerometers in each thigh and some straps to protect users from skin abrasion. This device weights 12 kg and is made of a composite of thermoplastic reinforced with aluminium [26].

8.5.2 Pneumatic Actuators

Pneumatic actuators use pressurised gas to generate output forces. These actuators have a simple and light structure being also relatively cheap [8]. They have a higher power-to-weight ratio than electric motors, but less than hydraulic actuators, providing a clean and non-flammable actuation method. However, they present low stiffness and accuracy, being used in a limited number of cases [8]. A good example of a pneumatically actuated exoskeleton system was proposed by Costa and Caldwell [19] for patients suffering from paralysis (Fig. 8.12a, b). This exoskeleton is formed by an aluminium, steel and carbon-fibre composite frame and has a total of 10 DOFs, 3 DOFs at both hips, 1 DOF at each knee and 1 DOF at each ankle. A potentiometer, mounted on the joint, measures the position, while the torque is measured by using an integral strain gauge [19].

One of the existing pneumatic actuated lower limb exoskeletons is the nurse robot suit as shown in Fig. 8.12c. This device was designed to augment the strength of a nurse by providing extra forces to carry patients without any back injuries. Whenever the nurse stands up, the robot suit helps in transferring the weight to the ground. It is also supported by actuators when the nurse bends the waist or knee [75]. The nurse robot suit consists of the shoulders, arms, waist and legs made of aluminium. The arms, waist and legs are driven by pneumatic rotary actuators [74].

8.5.3 Hydraulic Actuators

These actuators are rarely used in exoskeleton systems [30]. They usually use oil as pressurised fluid to transfer power to a joint, being able to quickly and precisely generate high torques [8, 30]. In comparison with the other actuators, the hydraulic ones present not only the highest power-to-weight ratio but also a complex structure requiring additional safety mechanisms due to the high power and stiffness [8]. Examples include the rehabilitation device designed by Kobetic et al. [40]. It was designed with a total of 6 DOFs with the hip joints being driven in flexion and extension by linear hydraulic actuators. An electromechanical system was considered to lock knee flexion and extension during the stance phase and unlock during the swing phase. The ankle joints (plantarflexion/dorsiflexion) were constrained to move in a sagittal plane [2].

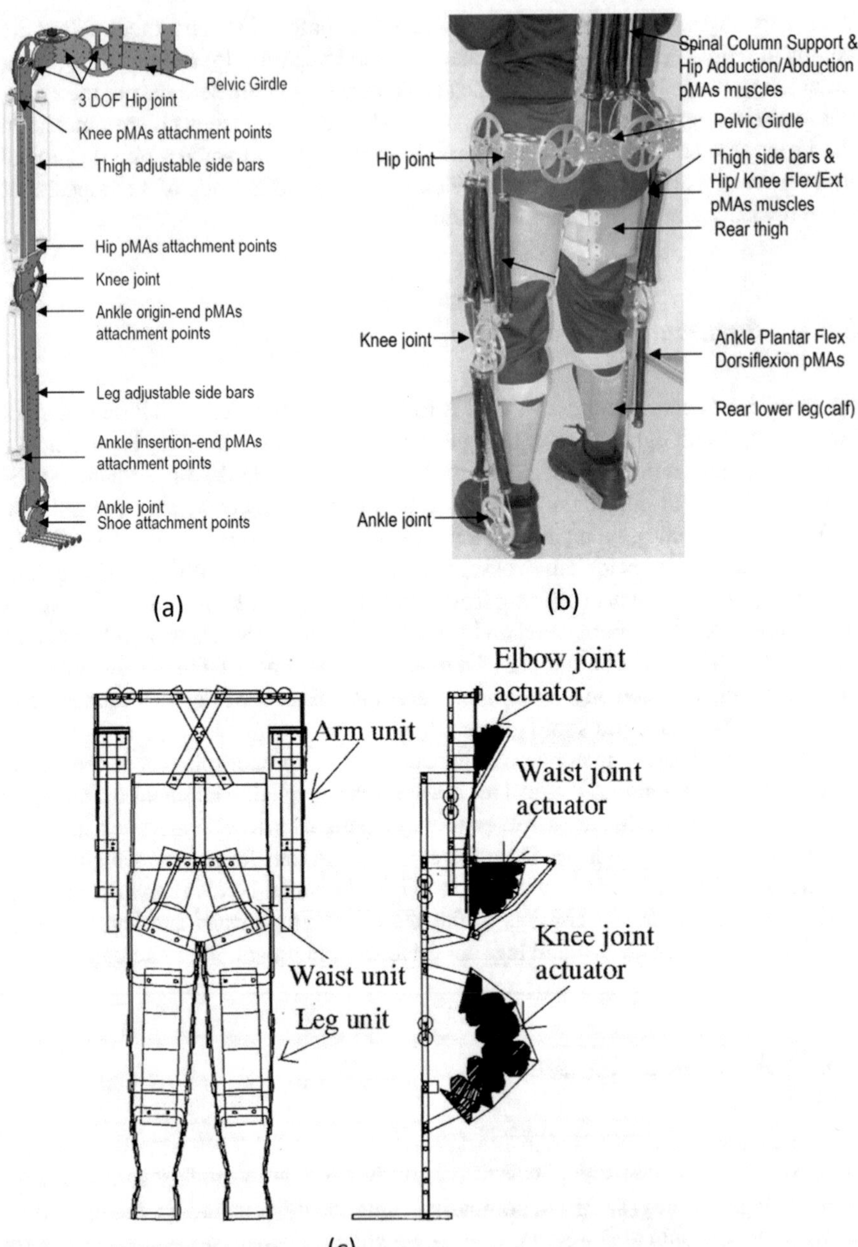

Fig. 8.12 (**a**) A representation of CAD model of the lower body exoskeleton. (**b**) The lower body exoskeleton [19]. (**c**) The nurse robot suit exoskeleton [74]

The Berkeley Lower Extremity Exoskeleton (BLEEX) is another example of an exoskeleton powered by hydraulic actuators [80] (Fig. 8.13). It is considered the first load bearing and energetically autonomous exoskeleton [38] and allows the

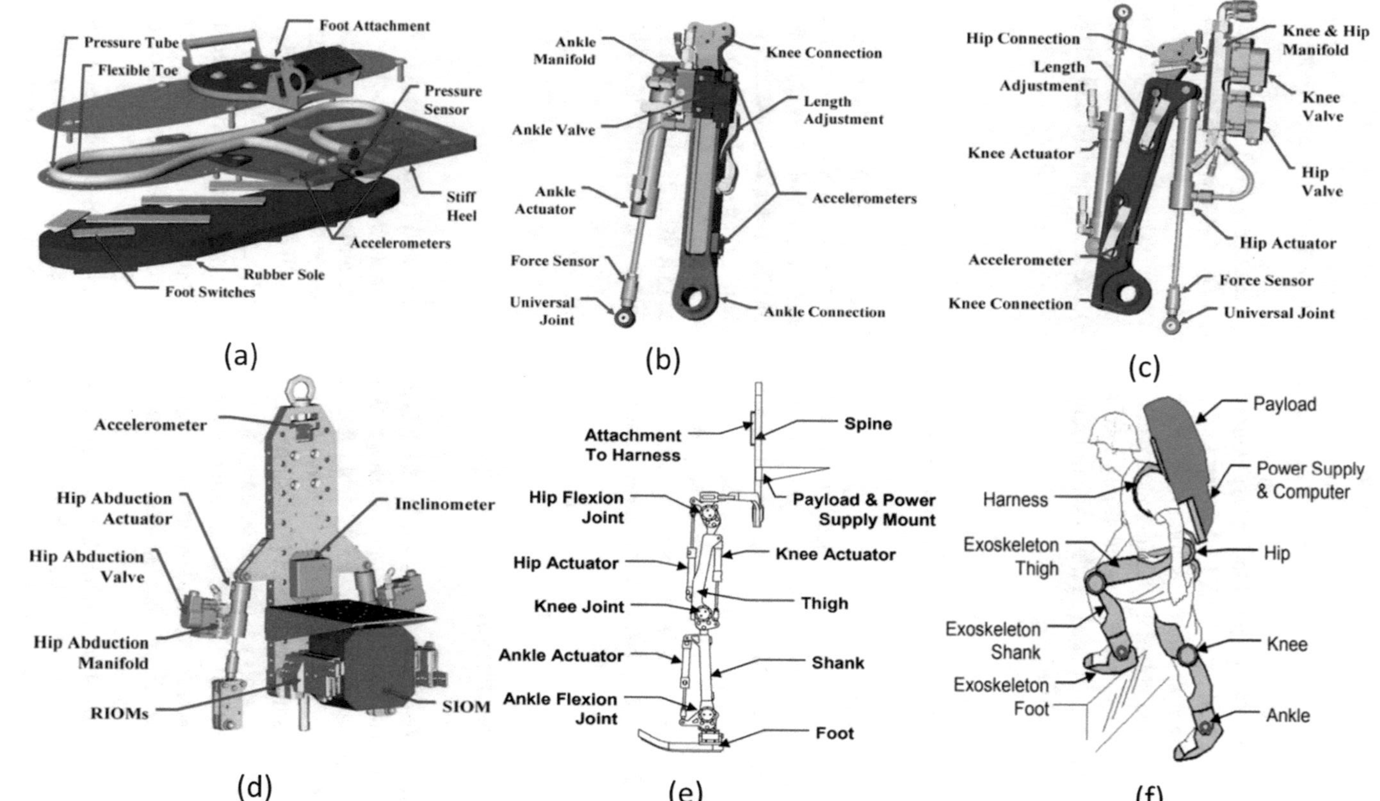

Fig. 8.13 (**a**) Foot design. (**b**) Shank design. (**c**) Thigh design. (**d**) Torso design. (**e**) Major components of BLEEX. (**f**) Final design of BLEEX [80]

user to carry a heavy load (34 kg) with minimum effort and walk with an average speed of 1.3 m/s [16, 75]. Moreover, it is also used to enhance the user's strength and endurance during locomotion [38]. The system was designed for emergency personnel such as soldiers, firefighter and disaster relief workers to carry significant payloads [16].

It consists of two powered legs, power unit and a backpack [16]. This exoskeleton has 7 DOFs per leg (3 DOFs at the hip joint, 1 DOF at the knee joint and 3 DOFs at the ankle joint). A total of 4 DOFs are active and powered by hydraulic actuators (ankle plantarflexion/ dorsiflexion, knee flexion/extension and hip adduction/abduction), while the remaining DOFs (hip internal/external rotation, ankle inversion/eversion and ankle adduction/abduction) use steel springs and elastomers [16, 75, 80].

8.6 Challenges and Opportunities

Although exoskeletons have been proven to be advantageous in many fields, there are several challenges that need to be addressed. Some of these challenges are related to the biomechanical design of the exoskeleton, safety and effective control algorithm. Additional challenges are related to the actuator selection, power supply and human/exoskeleton interface [30, 35]. According to Gopura et al. [30], the lack of the availability of proper accessory devices also constitutes a significant challenge for the exoskeleton's development. They stated that the actuation system and the technology of power transmission are not suitable enough to develop an ideal exoskeleton system.

8.6.1 Low Weight

In order to build portable exoskeletons, it is necessary to build exoskeletons as light as possible to improve their portability, making them more convenient for the user. It has been noted that materials of the frame structure and actuators are the main factors that affect the weight of these devices. Moreover, the fabrication method also has a significant impact [78]. Since these devices will become a second skin to the user, materials of these structures should be considered when developing these exoskeletons [30]. Thus, it is essential that these materials have some characteristics such as low density and toughness (e.g. the carbon fibre). Moreover, three-dimensional (3D) printing technology could be a great solution to create some components in order to obtain lightweight exoskeletons [12].

Overall, exoskeletons are heavy devices because the gearing systems and actuation units are not lightweight. However, exoskeletons should be relatively light to make the users feel comfortable when they wear the device [30]. In order to minimise the weight of the exoskeleton device, different materials and manufacturing technologies should be examined and tested for their effectiveness [61]. Lattice structures, presenting reduced weight and improved mechanical properties, will be more common. These structures could be designed and produced through a combination of topology optimisation and additive manufacturing. These lattice structures can also be used to improve the damping effects, increasing comfort.

8.6.2 Actuators

Actuators are one of the major challenges that need to be taken into account when developing robotic exoskeleton systems. For powered exoskeletons, specific features such as small volume, high power-to-weight ratio, high efficiency and compliance are necessary to be considered. The durability and lifetime of actuators also need to be improved [12]. New multifunctional actuators that integrate motor, brake and clutch functions into one device represent a good solution for improved exoskeletons [12]. The exoskeleton must also be designed taking into consideration the actuation system and its impact on the energy required to power the exoskeleton systems [11].

8.6.3 Cost

The cost of exoskeletons is an important issue influencing the development of lower extremity exoskeletons. Most users cannot afford these devices as they cost a significant amount of money [50]. A survey conducted by Wolff et al. [72] showed that the cost of the device constitutes the main concern for patients as it limits the adoption of exoskeletons in their daily lives. Therefore, researchers and engineers are required to make significant efforts to develop affordable devices. Prices are expected to decrease due to recent advances in robotics and mechatronic and also the reduction of the price of sensors and actuators [12].

Regarding rehabilitative exoskeletons, it can be said that there are several barriers that might limit the usage of this kind of exoskeletons in homes. Portability and cost are two factors that can affect the use of rehabilitation devices in patients' homes. There is also a general consensus that the exoskeleton devices are highly overpriced. However, the introduction of new technologies creates a possibility to design lightweight, cheap and portable exoskeletons [61].

8.6.4 Mechanical Design

Another challenge is related to the structure of the exoskeleton device that should have high strength and flexibility [30]. Furthermore, they must be adjustable to suit different users with different body weight and shapes. Modularity is in this case an important issue. These devices are required to have some features that allow the user to calibrate the exoskeleton to fit the wearer's requirements [35]. The mechanical design of some exoskeletons could reduce the performance of lower extremity exoskeletons and affect the biomechanics of the normal human gait. As a result, it will cause a discomfort to the user and increase the metabolic cost. In addition to this, the usage time of the device will be reduced. Moreover, some sections of the human body are complex to design. They require special designs to simulate the natural motion of the human [30]. Additionally, the mechanical structure of the device should be personalised to suit a particular individual. Another important consideration is that the noise caused by the exoskeleton should be minimised as it makes the wearer uncomfortable [12].

8.6.5 Safety

Safety is an important aspect of any exoskeleton devices and it should be given a great attention from both designers and manufacturers. As the use of an exoskeleton presents some risks (e.g. falls, fracture), further studies and efforts should be made to address them [33]. Furthermore, the safety of the battery needs to be considered when designing an exoskeleton. Shutdown systems in emergency conditions should be considered. Exoskeletons should have physical stops to limit the range of motion at joints, which must resist the maximum torque applied by actuators [12]. Dellon and Matsuoka [20] emphasise the importance of having safety standards for human–robot interaction, outlining some safety mechanisms that must be implemented such as limiting power output and limiting velocities of actuators. Since there is an interaction between the exoskeleton and the user, the exoskeleton devices are required to be mechanically compatible with the human anatomy, allowing the wearer to perform any movement safely and without any obstructions [11]. There are many factors that are critical and must be considered for ensuring user's safety, such as number of degrees of freedom, compliance with the measurements of the human body, range of motion, motion speed and the maximum force [32].

8.6.6 Human–Exoskeleton Interface

Young and Ferris [77] emphasise that the lack of understanding of the primary mechanisms that control the user's motion and the way that those mechanisms interact with the exoskeleton in parallel with the person is one of the major challenges in the field of exoskeleton research. Exchanging the information between the user and the device is one of the limitations of current exoskeletons. The reason is that some of the user's intentions cannot be quickly and accurately obtained by the sensors used in the devices. Therefore, new technologies such as artificial intelligence (AI) and neural technology will be extremely important for the development of future exoskeletons. Neural implants might be used to provide a feedback to the brain. Future exoskeletons will include EMG signals that predict the motion of the user and the neural implants as control systems. An electroencephalogram (EEG) can also be used to control the exoskeletons. An important point related to the human–robot interface is the occurrence of skin pressure sores. It has been found that the most common way to secure the device to the leg is by using the straps [11]. Therefore, these straps should be carefully designed to avoid skin issues [12].

Exoskeleton robots should not affect the functions of other body parts. The human–robot interaction (HRI) is significant when designing exoskeleton devices and should be customised to the individual's contour and anatomical needs [12]. Two aspects of HRI, namely, the physical human–robot interaction (pHRI) and the cognitive human–robot interaction (cHRI), should be considered. pHRI is related to the physical contact between the device and the user to transfer the power from the exoskeleton to the wearer or vice versa, while cHRI is related to the transmission of information from the user to the exoskeleton or vice versa [15].

8.6.7 Customisation and Personalisation

Exoskeletons must be designed and fabricated for the individual user. The main problem so far is associated with production costs. However, the emergence of additive manufacturing, a group of fabrication processes that create 3D objects by adding materials layer by layer contrary to convention subtractive processes, makes mass customisation and personalisation possible. Additive manufacturing (AM) comprises seven key technologies (see Table 8.2) (Vat photo-polymerisation, powder bed fusion, direct energy deposition, material extrusion, sheet lamination, material jetting and binder jetting) that allow the production of a wide range of metals, polymers, ceramics and composite materials. Key advantages of AM are as follows:

- Freedom of design: AM can produce an object of virtually any shape.
- Complexity for free: Increasing object complexity will increase production costs only marginally.

Table 8.2 Summary of all AM technologies

Technology	Principle
Vat photo-polymerisation	It is an additive manufacturing process in which a liquid in vat is selectively cured by light-activated photo-polymerisation
Powder bed fusion	It is an additive manufacturing that fuses regions of powder bed selectively through thermal energy
Direct energy deposition	It is an additive manufacturing process in which material is deposited from the nozzle and then melted by focused thermal energy
Material extrusion	Additive manufacturing processes in which material is melted and then extruded through a nozzle
Sheet lamination	It is an additive manufacturing process in which sheets of materials are bonded to form an object
Material jetting	It is one of the additive manufacturing processes that deposits wax and/or photopolymer droplets through a nozzle to create 3D object
Binder jetting	It is an additive manufacturing process that joins powder materials through the deposition of a liquid bonding agent

- Lightweight for free: Lightweight structures can be produced with reduced costs.
- No tooling required.
- Even complex objects are manufactured in one process step.
- Part consolidation: Reducing assembly requirements by consolidating parts into a single component, even complete assemblies with moving parts.

AM is also the most suitable technology for the production of small batches. The use of AM to produce certain parts of exoskeletons is now a reality, and we expect that this trend will significantly increase in the next years [4].

8.7 Conclusion

There is no doubt that lower extremity exoskeletons have significant roles in improving the quality of life of people with mobility disorders. These devices allow them to regain the ability to perform daily life activities. Exoskeletons have been designed for several purposes including rehabilitation, augmentation and locomotion assistance. They can be powered by some sensors and actuators, or they can be passive (unpowered). There are a number of exoskeletons already been commercialised for different purposes. However, this review identified areas for potential development. In order to produce exoskeleton with reduced weight and costs, and improved performance, it is important to use low-weight actuators and design devices with improved safety characteristics, improving also the mechanical design and the interface with users. The use of additive manufacturing will also contribute to the development of more personalised devices. Finally, artificial intelligence will also contribute to the development of smarter exoskeletons.

Acknowledgments The first author acknowledges the support received from the King Saud University to conduct his PhD studies.

Disclosure Statement No potential conflict of interest was reported by the authors.

Funding This research was funded by Saudi Arabian Government.

References

1. G. Aguirre-Ollinger, J.E. Colgate, M.A. Peshkin, A. Goswami, Active-impedance control of a lower-limb assistive exoskeleton, in *Rehabilitation Robotics, 2007. ICORR 2007. IEEE 10th International Conference on*, IEEE, June 2007, pp. 188–195
2. N. Aliman, R. Ramli, S.M. Haris, Design and development of lower limb exoskeletons: A survey. Robot. Auton. Syst. **95**, 102–116 (2017)
3. N. Aphiratsakun, M. Parnichkun, Balancing control of AIT leg exoskeleton using ZMP based FLC. Int. J. Adv. Robot. Syst. **6**(4), 34 (2009)
4. P. Balamurugan, G. Arumaikkannu, *Design for Customized Additive Manufactured Exoskeleton Using Bio CAD Modeling*, (2014) In: Proceedings of the International Conference on Innovations in Engineering and Technology (ICIET'14), 2014, Madurai, Tamil Nadu, India

5. S.K. Banala, S.K. Agrawal, J.P. Scholz, Active Leg Exoskeleton (ALEX) for gait rehabilitation of motor-impaired patients, in *Rehabilitation Robotics, 2007. ICORR 2007. IEEE 10th International Conference on*, IEEE, June 2007, pp. 401–407
6. M. Bouri, Y. Stauffer, C. Schmitt, Y. Allemand, S. Gnemmi, R. Clavel, P. Metrailler, R. Brodard, The WalkTrainer: A robotic system for walking rehabilitation, in *Robotics and Biomimetics, 2006. ROBIO'06. IEEE International Conference on*, IEEE, December 2006, pp. 1616–1621
7. E.B. Brokaw, I. Black, R.J. Holley, P.S. Lum, Hand Spring Operated Movement Enhancer (HandSOME): A portable, passive hand exoskeleton for stroke rehabilitation. IEEE Trans. Neural Syst. Rehabil. Eng. **19**(4), 391–399 (2011)
8. M. Brown, N. Tsagarakis, D.G. Caldwell, Exoskeletons for human force augmentation. Ind. Robot Int. J. **30**(6), 592–602 (2003)
9. R.S. Calabrò, A. Cacciola, F. Bertè, A. Manuli, A. Leo, A. Bramanti, A. Naro, D. Milardi, P. Bramanti, Robotic gait rehabilitation and substitution devices in neurological disorders: Where are we now? Neurol. Sci. **37**(4), 503–514 (2016)
10. P. Capodaglio, S. Vercelli, R. Colombo, E.M. Capodaglio, F. Franchignoni, Treadmills in rehabilitation medicine: Technical characteristics and selection criteria. G. Ital. Med. Lav. Ergon. **30**(2), 169–177 (2008)
11. M. Cenciarini, A.M. Dollar, Biomechanical considerations in the design of lower limb exoskeletons, in *Rehabilitation Robotics (ICORR), 2011 IEEE International Conference on*, IEEE, June 2011, pp. 1–6
12. B. Chen, H. Ma, L.Y. Qin, F. Gao, K.M. Chan, S.W. Law, L. Qin, W.H. Liao, Recent developments and challenges of lower extremity exoskeletons. J. Orthop Transl. **5**, 26–37 (2016)
13. B. Chen, C.H. Zhong, X. Zhao, H. Ma, X. Guan, X. Li, F.Y. Liang, J.C.Y. Cheng, L. Qin, S.W. Law, W.H. Liao, A wearable exoskeleton suit for motion assistance to paralysed patients. J. Orthop. Transl. **11**, 7–18 (2017)
14. F. Chen, Y. Yu, Y. Ge, Y. Fang, WPAL for human power assist during walking using dynamic equation, in *Mechatronics and Automation, 2009. ICMA 2009. International Conference on*, IEEE, August, 2009, pp. 1039–1043
15. G. Chen, C.K. Chan, Z. Guo, H. Yu, A review of lower extremity assistive robotic exoskeletons in rehabilitation therapy. Crit. Rev. Biomed. Eng. **41**(4–5) (2013)
16. A. Chu, H. Kazerooni, A. Zoss, On the biomimetic design of the Berkeley lower extremity exoskeleton (BLEEX), in *Robotics and Automation, 2005. ICRA 2005. Proceedings of the 2005 IEEE International Conference on*, IEEE, April 2005, pp. 4345–4352
17. S.H. Collins, M.B. Wiggin, G.S. Sawicki, Reducing the energy cost of human walking using an unpowered exoskeleton. Nature **522**(7555), 212 (2015)
18. J.L. Contreras-Vidal, N.A. Bhagat, J. Brantley, J.G. Cruz-Garza, Y. He, Q. Manley, S. Nakagome, K. Nathan, S.H. Tan, F. Zhu, J.L. Pons, Powered exoskeletons for bipedal locomotion after spinal cord injury. J. Neural Eng. **13**(3), 031001 (2016)
19. N. Costa, D.G. Caldwell, Control of a biomimetic "soft-actuated" 10dof lower body exoskeleton, in *Biomedical Robotics and Biomechatronics, 2006. BioRob 2006. The First IEEE/RAS-EMBS International Conference on*, IEEE, February 2006, pp. 495–501
20. B. Dellon, Y. Matsuoka, Prosthetics, exoskeletons, and rehabilitation [grand challenges of robotics]. IEEE Robot. Autom. Mag. **14**(1), 30–34 (2007)
21. I. Díaz, J.J. Gil, E. Sánchez, Lower-limb robotic rehabilitation: Literature review and challenges. J. Robot. **2011**, 1 (2011)
22. A.M. Dollar, H. Herr, Lower extremity exoskeletons and active orthoses: Challenges and state-of-the-art. IEEE Trans. Robot. **24**(1), 144–158 (2008)
23. M.A.M. Dzahir, S.I. Yamamoto, Recent trends in lower-limb robotic rehabilitation orthosis: Control scheme and strategy for pneumatic muscle actuated gait trainers. Robotics **3**(2), 120–148 (2014)
24. R. Ekkelenkamp, J. Veneman, van der H. Kooij, LOPES: Selective control of gait functions during the gait rehabilitation of CVA patients, in *Rehabilitation Robotics, 2005. ICORR 2005. 9th International Conference on*, IEEE, July 2005, pp. 361–364

25. A. Esquenazi, M. Talaty, A. Packel, M. Saulino, The ReWalk powered exoskeleton to restore ambulatory function to individuals with thoracic-level motor-complete spinal cord injury. Am. J. Phys. Med. Rehabil. **91**(11), 911–921 (2012)
26. R.J. Farris, H.A. Quintero, M. Goldfarb, Preliminary evaluation of a powered lower limb orthosis to aid walking in paraplegic individuals. IEEE Trans. Neural Syst. Rehabil. Eng. **19**(6), 652–659 (2011)
27. S. Freivogel, D. Schmalohr, J. Mehrholz, Improved walking ability and reduced therapeutic stress with an electromechanical gait device. J. Rehabil. Med. **41**(9), 734–739 (2009)
28. A.D. Gardner, J. Potgieter, F.K. Noble, A review of commercially available exoskeletons' capabilities, in *Mechatronics and Machine Vision in Practice (M2VIP), 2017 24th International Conference on*, IEEE, November 2017, pp. 1–5
29. I.D. Geonea, D. Tarnita, Design and evaluation of a new exoskeleton for gait rehabilitation. Mech. Sci. **8**(2), 307 (2017)
30. R.A.R.C. Gopura, K. Kiguchi, D.S.V. Bandara, A brief review on upper extremity robotic exoskeleton systems, in *Industrial and Information Systems (ICIIS), 2011 6th IEEE International Conference on*, IEEE, August 2011, pp. 346–351
31. J. Hamill, K. Knutzen, T. Derrick, *Biomechanical basis of human movement*, 4th edn. (Lippincott Williams & Wilkins, A Wolters Kluwer Business, 2015). https://doi.org/10.1017/CBO9781107415324.004
32. C. Hansen, F. Gosselin, K.B. Mansour, P. Devos, F. Marin, Design-validation of a hand exoskeleton using musculoskeletal modeling. Appl. Ergon. **68**, 283–288 (2018)
33. Y. He, D. Eguren, T.P. Luu, J.L. Contreras-Vidal, Risk management and regulations for lower limb medical exoskeletons: A review. Medical devices (Auckland, NZ), 10, p. 89, 2017
34. H. Herr, Exoskeletons and orthoses: Classification, design challenges and future directions. J. Neuroeng. Rehabil. **6**(1), 21 (2009)
35. Y.W. Hong, Y. King, W. Yeo, C. Ting, Y. Chuah, J. Lee, E.T. Chok, Lower extremity exoskeleton: Review and challenges surrounding the technology and its role in rehabilitation of lower limbs. Aust. J. Basic Appl. Sci. **7**(7), 520–524 (2013)
36. H. Kawamoto, Y. Sankai, Power assist system HAL-3 for gait disorder person, in *International Conference on Computers for Handicapped Persons*, Springer, Berlin, Heidelberg, July 2002, pp. 196–203
37. H. Kawamoto, S. Kanbe, Y. Sankai, Power assist method for HAL-3 estimating operator's intention based on motion information, in *Robot and human interactive communication, 2003. Proceedings. ROMAN 2003. The 12th IEEE international workshop on*, IEEE, October 2003, pp. 67–72
38. H. Kazerooni, J.L. Racine, L. Huang, R. Steger, On the control of the Berkeley lower extremity exoskeleton (BLEEX), in *Robotics and automation, 2005. ICRA 2005. Proceedings of the 2005 IEEE international conference on*, IEEE, April 2005, pp. 4353–4360
39. H. Kim, C. Seo, Y.J. Shin, J. Kim, Y.S. Kang, Locomotion control strategy of hydraulic lower extremity exoskeleton robot, in *Advanced Intelligent Mechatronics (AIM), 2015 IEEE International Conference on*, IEEE, July 2015, pp. 577–582
40. R. Kobetic, C.S. To, J.R. Schnellenberger, T.C. Bulea, R.G. CO, G. Pinault, Development of hybrid orthosis for standing, walking, and stair climbing after spinal cord injury. J. Rehabil. Res. Dev. **46**(3), 447 (2009)
41. S. Krut, M. Benoit, E. Dombre, F. Pierrot, Moonwalker, a lower limb exoskeleton able to sustain bodyweight using a passive force balancer, in *Robotics and Automation (ICRA), 2010 IEEE International Conference on*, IEEE, May 2010, pp. 2215–2220
42. D. Lim, W. Kim, H. Lee, H. Kim, K. Shin, T. Park, J. Lee, C. Han, Development of a lower extremity Exoskeleton Robot with a quasi-anthropomorphic design approach for load carriage, in *Intelligent Robots and Systems (IROS), 2015 IEEE/RSJ International Conference on*, IEEE, September 2015, pp. 5345–5350
43. K.H. Low, Robot-assisted gait rehabilitation: From exoskeletons to gait systems, in *Defense Science Research Conference and Expo (DSR), 2011*, IEEE, August 2011, pp. 1–10

44. K.H. Low, X. Liu, C.H. Goh, H. Yu, Locomotive control of a wearable lower exoskeleton for walking enhancement. J. Vib. Control. **12**(12), 1311–1336 (2006)
45. S. Marcheschi, F. Salsedo, M. Fontana, M. Bergamasco, Body extender: Whole body exoskeleton for human power augmentation, in *Robotics and Automation (ICRA), 2011 IEEE International Conference on*, IEEE, May 2011, pp. 611–616
46. A. Martínez, B. Lawson, M. Goldfarb, A controller for guiding leg movement during overground walking with a lower limb exoskeleton. IEEE Trans. Robot. **34**(1), 183–193 (2018)
47. M.M. Martins, C.P. Santos, A. Frizera-Neto, R. Ceres, Assistive mobility devices focusing on smart walkers: Classification and review. Robot. Auton. Syst. **60**(4), 548–562 (2012)
48. R.P. Matthew, E.J. Mica, W. Meinhold, J.A. Loeza, M. Tomizuka, R. Bajcsy, Introduction and initial exploration of an active/passive exoskeleton framework for portable assistance, in *Intelligent Robots and Systems (IROS), 2015 IEEE/RSJ International Conference on*, IEEE, September 2015, pp. 5351–5356
49. W. Meng, Q. Liu, Z. Zhou, Q. Ai, B. Sheng, S.S. Xie, Recent development of mechanisms and control strategies for robot-assisted lower limb rehabilitation. Mechatronics **31**, 132–145 (2015)
50. L. Mertz, The next generation of exoskeletons: Lighter, cheaper devices are in the works. IEEE Pulse **3**(4), 56–61 (2012)
51. A.K. Mishra, A. Srivastava, R.P. Tewari, R. Mathur, EMG analysis of lower limb muscles for developing robotic exoskeleton orthotic device. Procedia Eng. **41**, 32–36 (2012)
52. M. Mistry, P. Mohajerian, S. Schaal, An exoskeleton robot for human arm movement study, in *Intelligent Robots and Systems, 2005. (IROS 2005). 2005 IEEE/RSJ International Conference on*, IEEE, August 2005, pp. 4071–4076
53. S.M. Nacy, N.H. Ghaeb, M.M. Abdallh, A review of lower limb exoskeletons. Innovat. Syst. Des. Eng. **7**(11), 1–12 (2016)
54. D. Naidu, C. Cunniffe, R. Stopforth, G. Bright, S. Davrajh, Upper and lower exoskeleton limbs for assistive and rehabilitative applications, in *Presented at the 4th Robotics and Mechatronics Conference of South Africa (ROBMECH 2011)* (Vol. 23, p. 25), November 2011
55. Ü. Önen, F.M. Botsalı, M. Kalyoncu, M. Tınkır, N. Yılmaz, Y. Şahin, Design and actuator selection of a lower extremity exoskeleton. IEEE/ASME Trans. Mechatron. **19**(2), 623–632 (2014)
56. J. L. Pons, *Wearable Robots: Biomechatronic Exoskeletons* (Wiley, 2008), Chichester, England
57. M.B.I. Reaz, M.S. Hussain, F. Mohd-Yasin, Techniques of EMG signal analysis: Detection, processing, classification and applications. Biological Procedures Online **8**(1), 11 (2006)
58. D. J. Reinkensmeyer, V. Dietz (eds.), *Neurorehabilitation Technology* (Springer, 2016)
59. B.S. Rupal, S. Rafique, A. Singla, E. Singla, M. Isaksson, G.S. Virk, Lower-limb exoskeletons: Research trends and regulatory guidelines in medical and non-medical applications. Int. J. Adv. Robot. Syst. **14**(6), 1729881417743554 (2017)
60. B.S. Rupal, A. Singla, G.S. Virk, Lower limb exoskeletons: A brief review, in *Conference on Mechanical Engineering and Technology (COMET-2016)*, IIT (BHU), Varanasi, India, 2016, pp. 130–140
61. R. Soltani-Zarrin, A. Zeiaee, R. Langari, R. Tafreshi, Challenges and Opportunities in Exoskeleton-based Rehabilitation. arXiv preprint arXiv:1711.09523, 2017
62. K.A. Strausser, T.A. Swift, A.B. Zoss, H. Kazerooni, Prototype medical exoskeleton for paraplegic mobility: First experimental results, in *ASME 2010 Dynamic Systems and Control Conference*, American Society of Mechanical Engineers, January 2010, pp. 453–458
63. O. Unluhisarcikli, C. Mavroidis, P. Bonato, M. Pietrusisnki, B. Weinberg, Northeastern University, Boston and SPAULDING REHABILITATION HOSPITAL Corp, 2015. Lower extremity exoskeleton for gait retraining. U.S. Patent 9,198,821
64. O. Unluhisarcikli, M. Pietrusinski, B. Weinberg, P. Bonato, C. Mavroidis, Design and control of a robotic lower extremity exoskeleton for gait rehabilitation, in *Intelligent Robots and Systems (IROS), 2011 IEEE/RSJ International Conference on*, IEEE, September 2011, pp. 4893–4898

65. W. Van Dijk, H. Van der Kooij, XPED2: A passive exoskeleton with artificial tendons. IEEE Robot. Autom. Mag. **21**(4), 56–61 (2014)
66. W. Van Dijk, H. Van der Kooij, E. Hekman, A passive exoskeleton with artificial tendons: Design and experimental evaluation, in *Rehabilitation Robotics (ICORR), 2011 IEEE International Conference on*, IEEE, June 2011, pp. 1–6
67. C.L. Vaughan, B.L. Davis, C.O. Jeremy, *Dynamics of Human Gait*, 2nd edn. (1999). https://doi.org/10.1016/S0021-9290(01)00080-X
68. J.F. Veneman, R. Ekkelenkamp, R. Kruidhof, F.C. van der Helm, H. van der Kooij, A series elastic-and bowden-cable-based actuation system for use as torque actuator in exoskeleton-type robots. Int. J. Rob. Res. **25**(3), 261–281 (2006)
69. J.F. Veneman, R. Kruidhof, E.E. Hekman, R. Ekkelenkamp, E.H. Van Asseldonk, H. Van Der Kooij, Design and evaluation of the LOPES exoskeleton robot for interactive gait rehabilitation. IEEE Trans. Neural Syst. Rehabil. Eng. **15**(3), 379–386 (2007)
70. S. Viteckova, P. Kutilek, M. Jirina, Wearable lower limb robotics: A review. Biocybern. Biomed. Eng. **33**(2), 96–105 (2013)
71. K.P. Westlake, C. Patten, Pilot study of Lokomat versus manual-assisted treadmill training for locomotor recovery post-stroke. J. Neuroeng. Rehabil. **6**(1), 18 (2009)
72. J. Wolff, C. Parker, J. Borisoff, W.B. Mortenson, J. Mattie, A survey of stakeholder perspectives on exoskeleton technology. J. Neuroeng. Rehabil. **11**(1), 169 (2014)
73. S.K. Wu, M. Jordan, X. Shen, A pneumatically-actuated lower-limb orthosis, in *Engineering in Medicine and Biology Society, EMBC, 2011 Annual International Conference of the IEEE*, IEEE, August 2011, pp. 8126–8129
74. K. Yamamoto, K. Hyodo, M. Ishii, T. Matsuo, Development of power assisting suit for assisting nurse labor. JSME Int. J. Ser. C Mech. Syst. Mach. Elem. Manuf. **45**(3), 703–711 (2002)
75. T. Yan, M. Cempini, C.M. Oddo, N. Vitiello, Review of assistive strategies in powered lower-limb orthoses and exoskeletons. Robot. Auton. Syst. **64**, 120–136 (2015)
76. R. Yli-Peltola, *Lower Extremity Exoskeleton for Rehabilitation* (2017)
77. A.J. Young, D.P. Ferris, State of the art and future directions for lower limb robotic exoskeletons. IEEE Trans. Neural Syst. Rehabil. Eng. **25**(2), 171–182 (2017)
78. A. Zeiaee, R. Soltani-Zarrin, R. Langari, R. Tafreshi, Design and kinematic analysis of a novel upper limb exoskeleton for rehabilitation of stroke patients, in *Rehabilitation Robotics (ICORR), 2017 International Conference on*, IEEE, July 2017, pp. 759–764
79. L. Zhou, W. Meng, C.Z. Lu, Q. Liu, Q. Ai, S.Q. Xie, Bio-inspired design and iterative feedback tuning control of a wearable ankle rehabilitation robot. J. Comput. Inf. Sci. Eng. **16**(4), 041003 (2016)
80. A.B. Zoss, H. Kazerooni, A. Chu, Biomechanical design of the Berkeley lower extremity exoskeleton (BLEEX). IEEE/ASME Trans. Mechatron. **11**(2), 128–138 (2006)

Chapter 9
Bioglasses for Bone Tissue Engineering

Evangelos Daskalakis, Fengyuan Liu, Glen Cooper, Andrew Weightman, Bahattin Koç, Gordon Blunn, and Paulo Jorge Bártolo

9.1 Introduction

The bone is one of the hardest tissues of the human body and contributes to the composition of the human skeleton. The bones are responsible for supporting and protecting the organs and tissues of the body, producing white and red blood cells. Moreover, they provide shape and support to the whole body and help to the mobility and transmit of the forces that are produced during movement. The internal matrix of the bone is honeycomb-like that helps with the rigidity. The tissues of the bone are made from different cells such as osteocytes, osteoblasts and osteoclasts. Moreover, they are consisted of 10–20% elastic collagen fibers, that are also called ossein, 9–20% water and 60–70% minerals such as hydroxyapatite, cytokines, growth factors, sodium, bone sialoprotein, osteonectin, magnesium, carbonate, chondroitin sulfate, potassium, osteocalcin, phospholipids, hyaluronic acid, osteopontin, calcium phosphate, and phosphoproteins that are very important for the regeneration and remodeling of the bone [14, 77].

The functionality of the human body can be interrupted from defects or fractures that occur in the human skeleton. These abnormalities in the skeletal system

E. Daskalakis · F. Liu · G. Cooper · A. Weightman · P. J. Bártolo (✉)
School of Mechanical, Aerospace and Civil Engineering, University of Manchester, Manchester, UK
e-mail: evangelos.daskalakis@postgrad.manchester.ac.uk; Fengyuan.liu@manchester.ac.uk; glen.cooper@manchester.ac.uk; andrew.weightman@manchester.ac.uk; paulojorge.dasilvabartolo@manchester.ac.uk

B. Koç
Faculty of Engineering and Natural Sciences, Sabanci University, Istanbul, Turkey
e-mail: bahattinkoc@sabanciuniv.edu

G. Blunn
School of Pharmacy & Biomedical Sciences, University of Portsmouth, Portsmouth, UK
e-mail: gordon.blunn@port.ac.uk

P. J. Bártolo, B. Bidanda (eds.), *Bio-Materials and Prototyping Applications in Medicine*, https://doi.org/10.1007/978-3-030-35876-1_9

can be caused by many factors such as osteoarthritis, tumors, resection, fall of bone's density, metabolic diseases, and trauma. In these cases, the human body is unable to regenerate and produce new bone cells and materials, that leads to greater bone resorption than bone deposition. If the bone lesion is at a critical stage and size, the cure is hardly manageable, and in most cases, the nonunion of the bone is inevitable. For the treatment of the most defects, biocompatible and biodegradable implants are used in order to help and guide the cells to proliferate and differentiate into bone tissue.

Bone defects can be treated using a wide range of biological grafts such as autografts, allografts, and xenografts. On the one hand, autograft is the procedure when the skin is transplanted to cover large full-thickness skin lesions, based on the transplanting of split-thickness grafts. Allografts are grafts that have been removed from other people and are used to prevent the loss of fluids, to reduce the pain, to promote the healing of tissues that are under the epidermis and to prevent the infection of the wound. However, autografts are limited in use, because they produce scars at the donors and their hospitalization time is increased. On the other hand, allografts present some problems related to religious issues associated to their use and problem-associated disease transmission and rejection by the immune system [26, 62]. As stated by Pereira and Bártolo [62], xenografts are grafts that can be taken from one person and implanted to another person that are not the same species. Important advantages of xenografts include high availability and low cost [62]. However, they present also some limitations, namely, the risk of rejection, disease transmission, lack of efficient skin regeneration, and ethical problems [26, 62].

To solve these problems, the concept of synthetic grafts (scaffolds) made from a wide range of biocompatible and biodegradable materials has been investigated. These scaffolds are usually made of polymers, ceramics or polymer–ceramic materials. Recently bioglass emerged as a material with high potential for the fabrication of scaffolds. Bioglass materials are biocompatible, osteoconductive, biodegradable, and osteoinductive, that makes them suitable for biological uses, because of their fast surface reaction with the fluids of the body and good bond creation with soft and tissue (Hench and Julia 2002).

This chapter discusses key characteristics of bioglass materials and presents the main technologies being used to weak scaffolds incorporating bioglass materials.

9.2 Bioactive Glasses

In 1969, Larry L. Hence discovered an amorphous solid silica-based material, which is biocompatible and biocompatible, with the ability to bond on soft and hard tissue [38]. These bioactive glasses react with the physiological fluids forming a layer of crystallized hydroxycarbonate apatite (HCA) between the glass and the bone. The structure and composition of that layer is similar to that of the bone mineral, which leads to the formation of a strong bond without the need of fibrous tissue [4, 74, 81]. The composition of bioactive glasses is very specific. They contain

45 wt% of SiO_2, 24.5 wt% of Na_2O, 24.5 wt% of Cao and 6 wt% of P_2O_5, and because of their osteogenic, osteoconductive, and osteoproductive abilities, they are categorized as biomaterials of class A compared with the biomaterials of class B that are only osteoconductive, such as HA [38]. On the surface of bioactive glasses, soluble ions of Ca, Si, Na, and P are released, because of reactions, causing intracellular and extracellular responses, that lead to the formation of physiological solutions such as HA [16, 37, 75].

However, the layer of HCA formation and the reaction rate of the bioactive glasses is very important for their osteogenesis and bioactivity. Although bioactive glasses are considered biodegradable materials, their composition and surface area, that depends on the morphology and the particle size for the formation of the scaffolds, affect the degradation rate [46]. For that reason, Hence showed that small changes in the composition of the bioactive glass can define if they are bioactive, bioinert or resorbable (Fig. 9.1).

From Fig. 9.1, it is possible to identify the different compositional regions (A, B, C, D and E). In region A, the bioactive glasses have a constant of 6 wt% of P_2O_5, and the compositions are bond to bone and bioactive. In region B, the compositions are bioinert, resulting to the formation of capsules with nonadherent fibrous. In region C, the compositions are resorbable. Region D is formed because of technical factors. In region E, the tissue bonding is soft [38].

Bioactive glasses of class A have 11 stages of reaction at the surface (Fig. 9.2), where dissolution, exchange of ions and condensation take place on the surface of

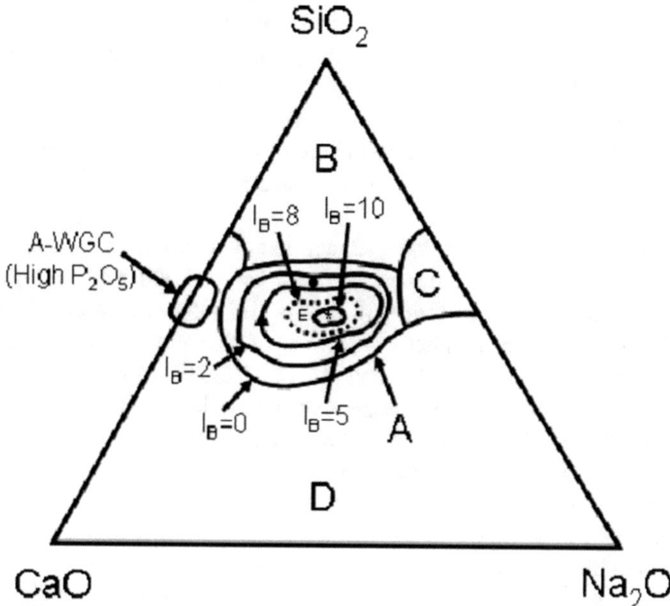

Fig. 9.1 Compositional dependence between the bone and soft tissue from the use of glass ceramics and bioactive glasses [39]

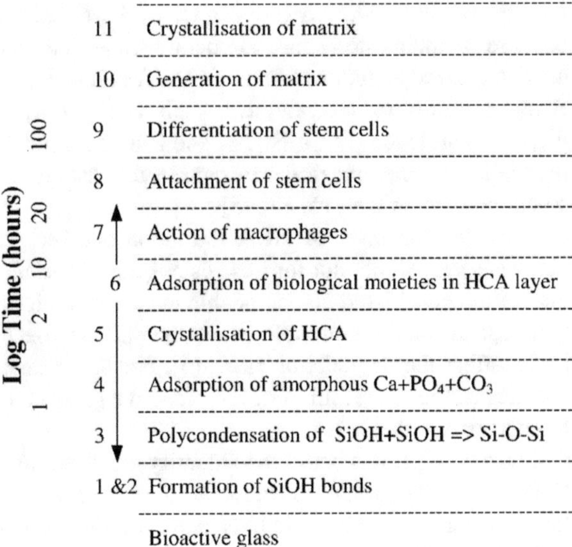

Fig. 9.2 Sequence of the involved reactions during the bond creation between the bioactive glass and the tissue [38, 42]

the glass. Then a pre-layer of HCA is created. The reaction on the surface of the glass increases the proliferation, colonization and differentiation of the cells [32, 42].

The oldest composition of bioactive glass, which is called 45S5 and was invented on 1971, is consisted from a silica-based network Na_2O–CaO–P_2O_5–SiO_2. A melt-derived process was used for the production of 45S5 bioglass, with highest content of 60 wt% SiO_2. Moreover, different bioactive glasses were produced, such as CaO–P2O5–SiO2 with the use of sol–gel technique, making them more bioactive and can be absorbed faster because of the larger surface area resulting from their porosity of nanometer texture (Jones et al., [47, 48]). With the use of a sol–gel technique, the content range of SiO_2 can be increased from 60 wt% to 85 wt%. Moreover, Na_2O has been characterized as a non-critical component for the composition of the bioglass [73].

Furthermore, a lot of studies have shown that the products of bioactive glasses with ionic dissolution increase the osteogenesis by controlling the cell differentiation, proliferation, production of growth factors and gene expression, showing that the rate of bone formation is higher with the use of bioactive glasses compared with the inorganic ceramics such as TCP and HA [61, 78, 79]. Moreover, other studies showed that bioactive glasses without phosphate have higher potentials for the production of tissue scaffolds than those with phosphate. This happens because the ions of the phosphate exist as orthophosphate ions instead of forming a vitreous network [49]. The connectivity of the glass network can be increased, with the use of phosphate, with the isolation of the calcium cations. This will have as a result the decrease of the bioactivity and the dissolution rate of the glass [46].

Other investigations have tried concentrations of SiO_2 up to 90 wt% with the use of sol–gel technique and characterized the bioactivity of the glasses [55]. On the one hand, Hench and Saravanapavan proved that $70SiO_2$–$30CaO$ system had the same bioactivity with 45S5 or 58S melt-derived bioglass and thought to be the basis for a lot of third-generation tissue engineering materials [41, 65].

On the other hand, bioactive glasses have some disadvantages such as low values of fracture toughness and very low mechanical strength. This has as a result the limited use of bioactive glasses in situations of load bearing [80]. One significant modification that led to the improvement of the Ceravital, Bioverit, and apatite–wollastonite (A–W) glass-ceramics which are bioactive materials under A class was the formation of crystalline particles with heat treatment. The production of A–W glass-ceramics happens with the partial crystallization of the glass matrix, using high temperature (870 °C–900 °C) for a long time, producing nucleation of crystallization [21, 22]. Moreover, A–W glass-ceramics have higher bending strength compared with the bioactive glasses because of the reinforced apatite phases from β-wollastonite, increasing the fracture toughness making it a better choice for bone regeneration [9, 80]. The crystalline microstructures can prevent the straight spreading of the cracks, leading them to branch out or deflect and to increase the resistance to the crack [80].

9.3 Scaffold Design and Fabrication

9.3.1 Design Requirements

Once an implantation of a scaffold takes place, the materials have to degrade in a rate that is suitable that will match with the rate that the new tissue grows around the scaffold [13]. For that reason, in order to consider a material suitable for tissue engineering applications, it has to degrade in a controlled rate [13]. Moreover, the degradation must not produce inflammatory response and has to be nontoxic [52].

Furthermore, a suitable material for tissue engineering applications has to facilitate cell proliferation, differentiation and, most importantly, attachment ([51]; Liu et al. 2012). The by-products of the material have to be nontoxic, and after the implantation, it must not activate the inflammatory response of the immune system [51, 60]. For that reason, the material's chemical make-up is very important for the appropriate behavior of the material after the implantation. While these materials can be biodegradable and biocompatible by their own, if they are combined with other materials, for the formation of an implant device, they might be no longer biocompatible or biodegradable [7, 64]. The material has to offer a biocompatible surface where the growth of a new bone will be possible (osteoconductive) [1, 3, 15]. Moreover, it has to promote differentiation of the stem (osteoinductive) [1, 3, 15]. Finally, the materials should cause the growth of a new bone by inducing extracellular and intracellular responses between the tissue and the material (osteoproductive) [19, 36, 72].

9.3.2 Structural Properties

The mechanical properties are also very important for the materials due to the need of support during the regeneration of the bone tissue [82]. The material must have mechanical properties such as compression strength, elastic modulus and fracture toughness, that should mimic the mechanical properties of the tissue around the implant [33, 56]. The repair of the bone includes interrelated mechanical, biological and chemical processes that happen simultaneously, where the mechanical conditions are responsible for the correct function of many biological and chemical processes [5, 23, 35]. The reduction of the bone tissue loss because of the stress-shielding can be occurred with the production of material where it's mechanical properties will mimic those of the tissue [54, 68]. Except from the mechanical properties, there are some others that are required for a suitable scaffold. These properties are the surface roughness, the porosity, and the wetting ability (Fig. 9.3).

9.3.3 Porosity

Porosity is one of the most important parameters for the fabrication of a scaffold because it defines how much applicable is the scaffold in tissue engineering. Porosity is a very important parameter because of its different functions (Fig. 9.4) [11, 12]. The pore parameters such as interconnectivity, morphology, and orientation have to adapt in order to respond to the needs of the tissue. More specifically, the fabricated scaffold must have interconnected pores, and the pore size should be between 200–500 μm [63]. Moreover, the fabricated scaffolds should be produced in a way

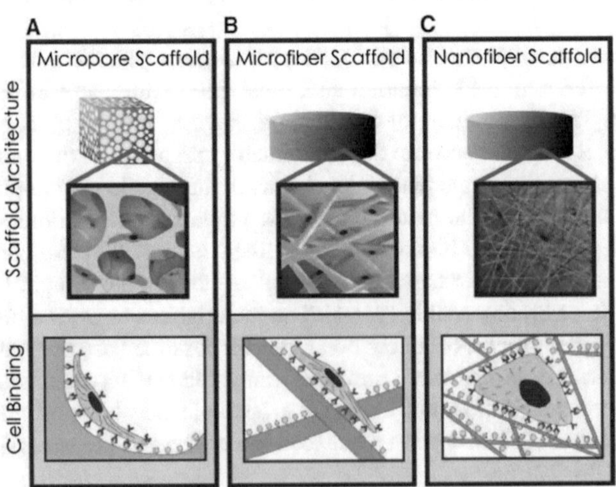

Fig. 9.3 Behavior of the cells on different surfaces (Stevens and George 2005)

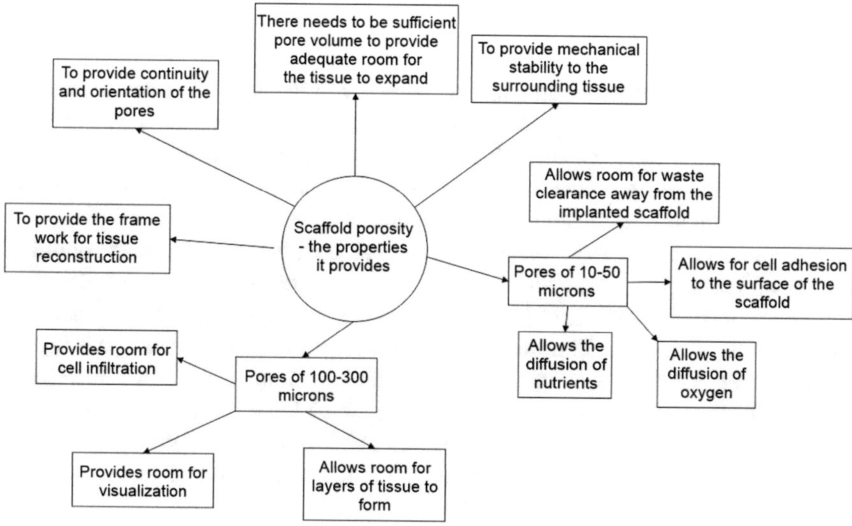

Fig. 9.4 Different functions related to the scaffold's pore structure

to maximize the required porosity but to keep, at the same time, its mechanical properties [21, 22, 50]. A human trabecular bone has a porosity of 50–90% [11, 12, 30, 66, 69], meaning that the scaffolds should have the same porosity in order to connect the scaffold with the neighbor tissue [50]. The scaffold's porosity can be calculated with the following mathematical equation:

$$\text{Porosity} = 1 - \left(\frac{\rho_{\text{scaffold}}}{\rho_{\text{material}}} \right) = 1 - \left(\frac{m_{\text{scaffold}}}{v_{\text{scaffold}} \times \rho_{\text{material}}} \right), \tag{9.1}$$

where ρ_{scaffold} and ρ_{material} are the scaffold's densities and material from which it is fabricated, respectively, m_{scaffold} is the scaffold's mass, and ν_{scaffold} is the scaffold's total volume.

9.4 Surface Topography

The topography of the surface is also one very important parameter for the use of a scaffold in bone tissue engineering, and it is related with the contact angle [43]. The cell attachment on the tissue and the contact angle are affected directly by any changes in the roughness of the surface [43]. Moreover, osteoblasts can be sensitive to the roughness of the surface, and sometimes the roughness of the surface affects negatively the healing time (Nasatzky et al. 1993). The surface of the scaffold should have a specific roughness for cells to attach and proliferate (Anselme et al. 2000).

9.5 Wettability

A liquid has the ability to maintain the contact with the surface of a solid sample. This ability is called wetting. The value of wetting can be calculated by a balance force between the cohesive and adhesive forces [17, 57]. Wetting is very important for a biomaterial because when it comes in contact with a tissue, it can determine the cell attachment and the protein absorption [24].

For a cell to attach on a biomaterial, the value of the contact angle should be between 45° and 90°. This means that the surface of the biomaterial is hydrophilic and allows the cell attachment [24, 43]. If the contact angle is lower, it means that the nature of the material is more hydrophilic, whereas if the contact angle is higher, it means that the nature of the material is more hydrophobic [10]. The relation between the liquid droplet and the acting of the interfacial tension on the surface and the relation between the capabilities of wetting and the contact angle on the surface are shown in Fig. 9.5. The equations used for the calculation of the equilibrium contact angle (θ_c) are:

$$\gamma_{LV}\cos\theta = \gamma_{SV} - \gamma_{SL}, \tag{9.2}$$

where γ_{LV}, γ_{SV}, and γ_{SL} are the surface tension for liquid–vapor, the surface energy of the polymer, and the solid–liquid surface tension, respectively.

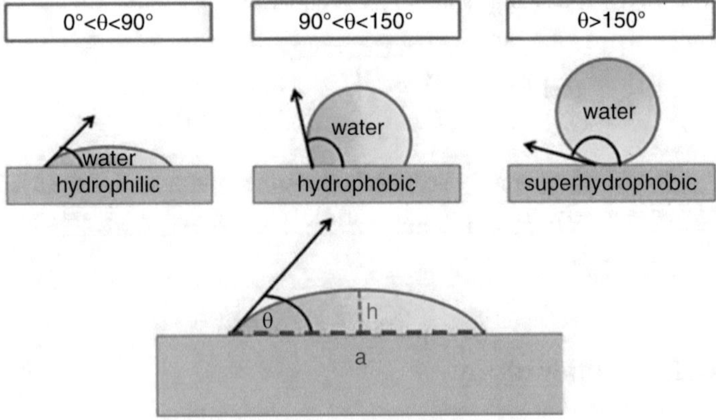

Fig. 9.5 Contact angle of sample's wettability and the hydrophobic or hydrophilic nature and the wettability relationship of the sample (Mattone et al. 2017)

9.6 Manufacturing Processes for Scaffolds Fabrication

Different techniques such as foam replication and sol–gel processing have been incorporating bioglass.

9.6.1 Foam Replication

The foam replication is a technique where a foam template is dipping ranging from a foam of polyurethane to coral or wood to a glass-ceramic contained in a ceramic suspension and then at a binder [20] (Fig. 9.6). On the one hand, foam replication is used for the production of high-porous 3D scaffolds where the size and shape of the fabricated scaffolds can be managed easily and can be reproduced. In the case where the foam template is a polyurethane, the fabricated scaffolds obtain a porous network because of polyurethane's nature (Boccaccini et al. 2007; [58]). On the other hand, for the high porosity that foam replication gives to the scaffolds, which in some case is 90%, the mechanical strength is lower compared with other techniques. This problem can be resolved by adding coatings of polymer [20], leading to the increase of compression strength without changing the porosity of the scaffold.

9.6.2 Sol–Gel Processing

The sol–gel processing is a technique where the used material homogeneity can be controlled along with its chemical composition (Figs. 9.7 and 9.8). This technique involves gel foaming with the help of a surfactant [31]. Then, the solution is going through a condensation and gelation process, at room temperature. Then it is dried and heated, forming a glass for the increase of structure's strength. Finally, the product is sintered after the removal of the liquids. The scaffolds produced from this technique have high bioactivity and high porosity, having mesopores and macropores. However, these scaffolds have low mechanical strength. This problem can be solved by polymer coatings and composite scaffolds [31].

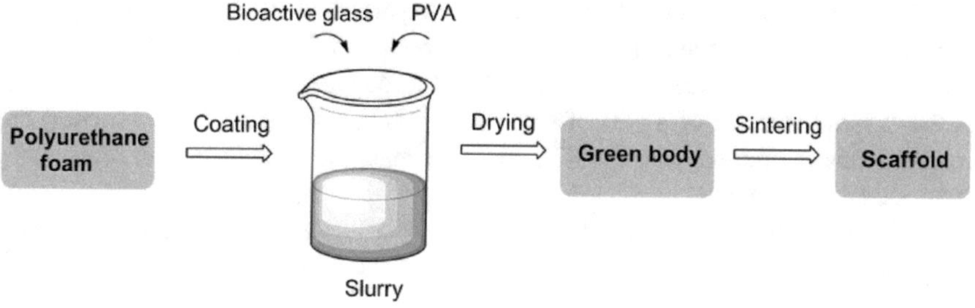

Fig. 9.6 Foam replication technique process [44]

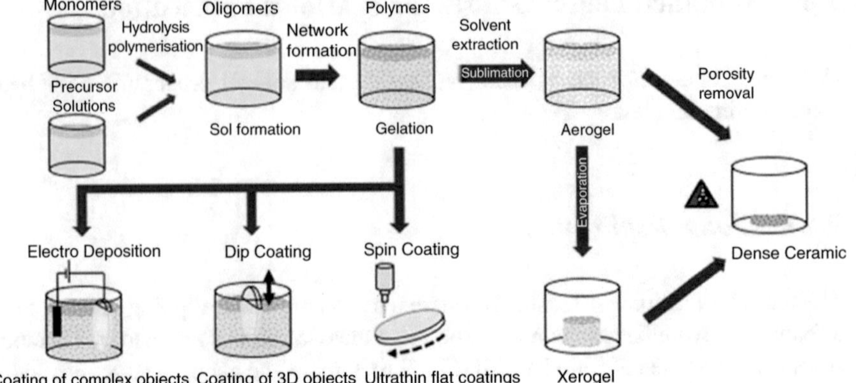

Fig. 9.7 Sol–gel processing technique [59]

9.7 Additive Manufacturing

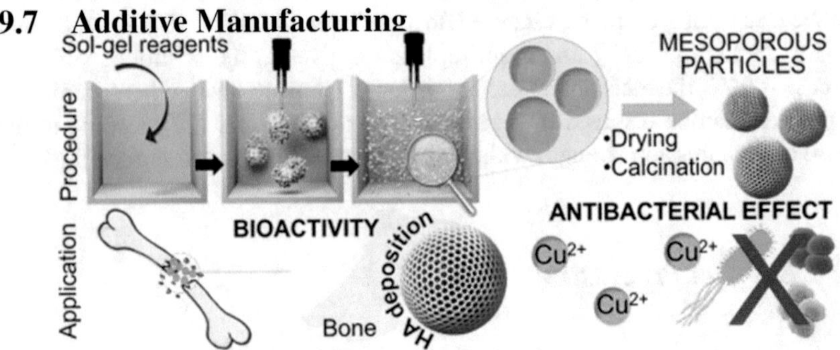

Fig. 9.8 Sol–gel technique procedure and application [6]

Conventional techniques mentioned before are simple but require significant human information and present limited control over pore shape, pore size, and pore interconnectivity. Additive manufacturing emerged as a group of technologies that solve some of these limitations. Through additive manufacturing, 3D scaffolds are created by adding material layer by layer.

First, a solid model of the 3D scaffold should be generated in a computer. Then, a STL file is created, which is the standard file format of the input data for all the processes of additive manufacturing. The 3D model, that has been saved in the STL file, is represented by several triangles, called three-sided planar facets. Each facet is the external surface of a part of the model. Then, thin layers (sliced model) are mathematically created from the STL model. Finally, a suitable system reproduces the sliced data (Fig. 9.9).

Different techniques such as ink-jet printing, extrusion-based processes, powder-bed fusion, and vat photopolymerization have been exploited to produce bone scaffolds incorporating bioglass.

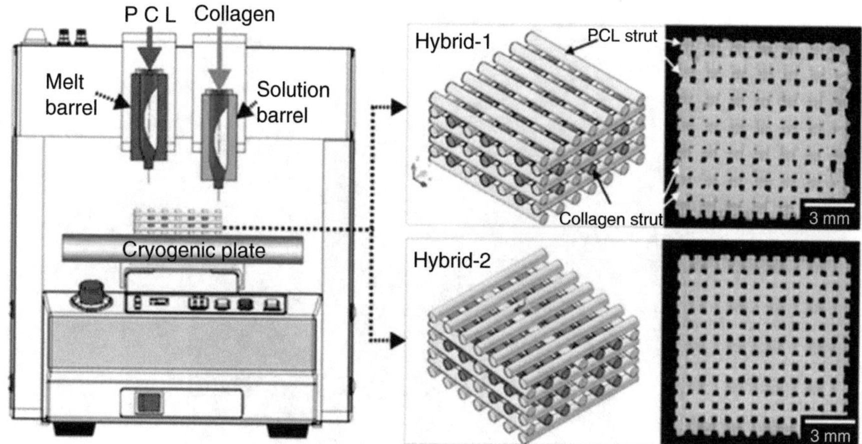

Fig. 9.9 Solid free-form fabrication technique process [2]

9.8 Ink-Jet Printing Process

Ink-jet printing process is an additive manufacturing technique deposition of a material conservation used for materials in liquid phase. These materials are consisted by a dissolved solute or a dispersed solute in a solvent [67]. The process of this technique involves the production of fluid droplets, called inks. The creation of an image is happening with colorants inside the ink. The ink-jet printing process technique is an evolved technology used for medical and industrial applications (Wijshoff 2017). Along with the development and evolution of applications, the science behind the ink-jet printing also developed, containing the manipulation of fluids in small amounts (Wijshoff 2017).

The process of ink-jet printing involves an ink of fixed quantity inside a chamber, which is ejecting from a nozzle, with the use of a piezoelectric action by reducing suddenly the volume of the chamber. A chamber, with a liquid inside, is contracted in reply to the voltage. The reduction sends a shockwave inside the liquid, causing the ejection of a liquid drop from the nozzle (Wijshoff 2017; [67]). The liquid drop because of the air resistance and gravity falls until it hits the substrate, and then it spreads, due to the motion, under momentum, and it flows on the surface due to the surface tension (Fig. 9.10). Then the droplet dries due to solvent evaporation (Fig. 9.11). The viscosity is an important parameter that affects the spreading and the final printed shape of the drop, which is a function of the polymer molar mass (Wijshoff 2017; [67]). Moreover, the height of the printing head affects severely the shape of the drop, as a function of polymers concentration.

Elsayed et al. used an ink for 3D printing. At the beginning, MK silicone was dissolved inside of isopropanol alcohol of 27.5 vol.% and mixed with fume silica, which had a nano-size of 7 nm and a surface area of 220–280 m^2.gr. Silica was used for the modification of rheological properties, in order to give a pseudo-plastic

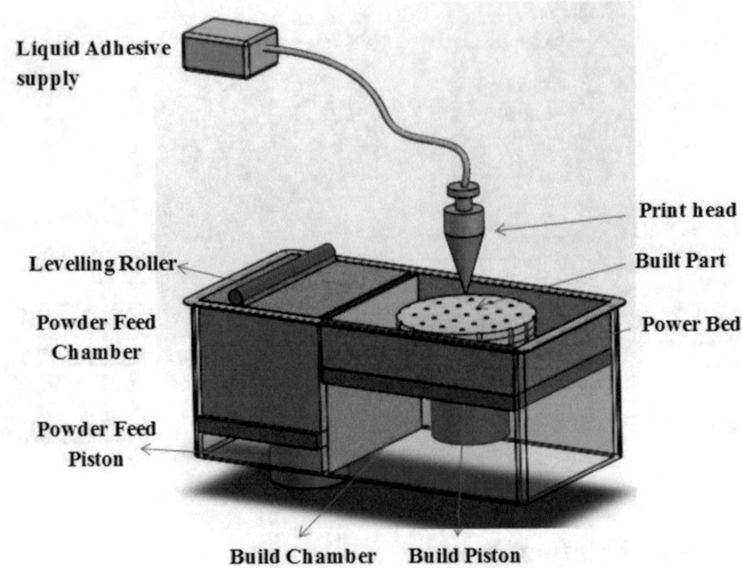

Fig. 9.10 Ink-jet printing process (Liu et al. 2018)

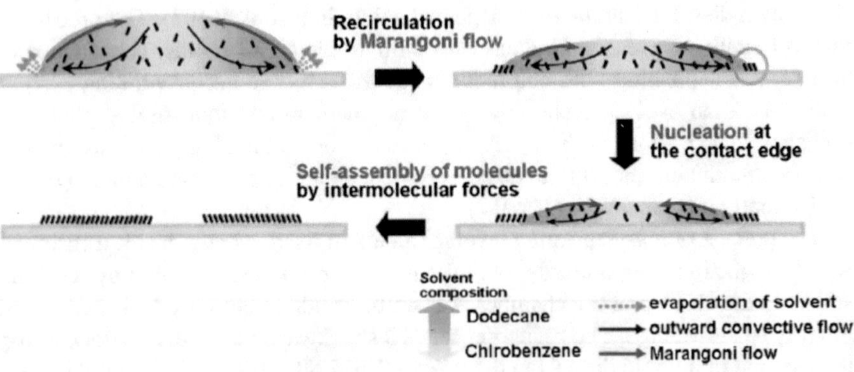

Fig. 9.11 Drying process of the drop after its deposition through the ink-jet printing process (Singh 2010)

behavior, which is suitable for 3D printing [29]. Moreover, silicone–fumed silica–$CaCO_3$, Na_2CO_3–Na_2HPO_4 proportions were adapted in order to keep constant the oxide formulation. With the use of a planetary ball mill and an alumina jar, the solution was mixed at 150 rpm for 1 hour, which led to the creation of a silicone-fumed silica gel. The created gel had no agglomerate [29]. Then, fillers of Biosilicate glass-ceramic ($CaCO_3$, Na_2CO_3, and Na_2HPO_4) were mixed with the gel solution creating an ink. The ink was, then, placed inside a plastic syringe for ink writing. Then, pasta filaments were produced, layer by layer, with the use of a printer (Fig. 9.12). The diameter of the nozzle was 0.81 mm, a porous size of 600 μm, porosity of 60%, and a mean compression strength of 6.7 MPa [29].

9.8.1 Extrusion-Based Process

Extrusion-based processes is a simple and cheap additive manufacturing technique based on a three-axis stage of motion for the creation layer by layer of complex structural forms (Lacey et al. 2018) (Fig. 9.13). Extrusion-based process is a technique friendly to the cells and allows the deposition of several materials simultaneously, making it the most preferable choice for bioprinting (Serex et al. 2018).

Extrusion-based process is a set of technique where a screw, pneumatic system, or plunger uses pressure to push cell suspensions and material through a small syringe (nozzle). During the extrusion from the nozzle, the material lays down and forms a scaffold, providing physical support holding the cell suspension at its place (Chartrain 2018) (Fig. 9.14). The print extrusion heads are very popular because they can print materials and cell types with a variety of cell densities and viscosities (Chartrain 2018; Liu et al. 2018). Nevertheless, due to extrusion velocities that can be very high and the diameter of the nozzle that can be very small, the cells can be affected from huge shear forces (Chartrain 2018). These shear forces can destroy the membranes of the cells and have an enormous effect on cell's viability, which continues even after the end of the process. With the increase of nozzle's diameter, the shear forces can be reduced but will affect the resolution but does so at the expense of resolution.

A method of extrusion-based process called freeze extrusion fabrication, which is a combination of freeze-drying with extrusion printing, was used for the fabrication of 13–93 glass 3D scaffolds [27, 45, 46]. During this process, a bioactive polymer-glass paste was deposited, through extrusion, layer by layer at cold environment. The water was removed with the use of freeze-drying before the sintering of the paste at 700 °C. The scaffolds had a porosity of 50%, porous size of 300 μm, and compression strength of 140 MPa [27, 45, 46] (Figs. 9.15 and 9.16).

9.8.2 Powder Bed Fusion Process

Powder bed fusion (LPBF) is a process where a laser heats and melts selectively layers of metal materials in powder form, producing complex parts of three dimensions, with the use of computer-controlling laser beam of high-energy in a secure atmosphere (Liu et al. 2018; Robinson et al. 2018). The laser beam tracks each cross-section's shape of the building model, by sintering, inside a thin layer, metal powder (Liu et al. 2018; Robinson et al. 2018). Moreover, during the process, energy is supplied, fusing the surrounding particles of the powder and bonding the new

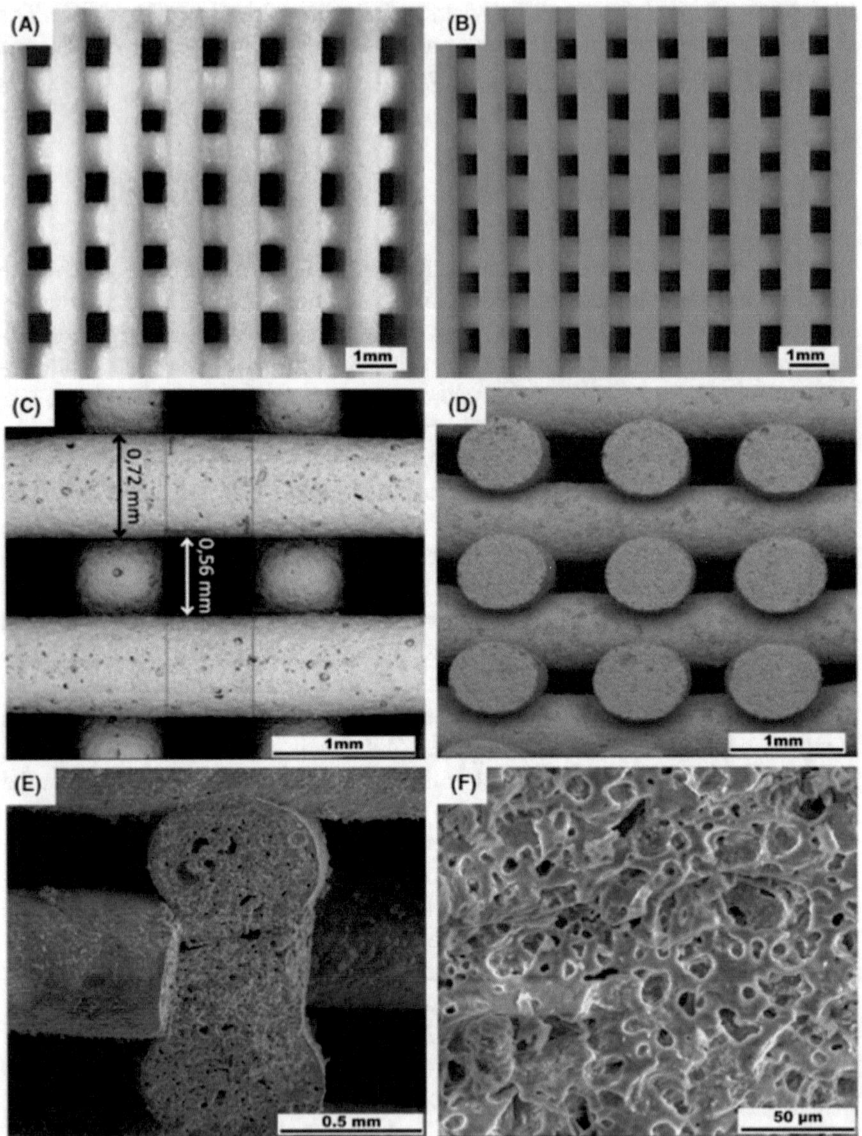

Fig. 9.12 SEM images of 3D scaffolds. (**a**) View of top surface before firing. (**b**) View of top surface of ceramized scaffold. (**c**) Spacing of layers after ceramization, (**d**), (**e**), and (**f**) Details of cross section [29]

layer with the previously sintered. After the solidification of each layer's, the piston above the building model is withdrawing to a new position where a mechanical roller supplies the piston with a new layer of metal powder (Liu et al. 2018). The remaining powder that is not affected from the beam laser works as a support for the building model until the end of the process (Fig. 9.17).

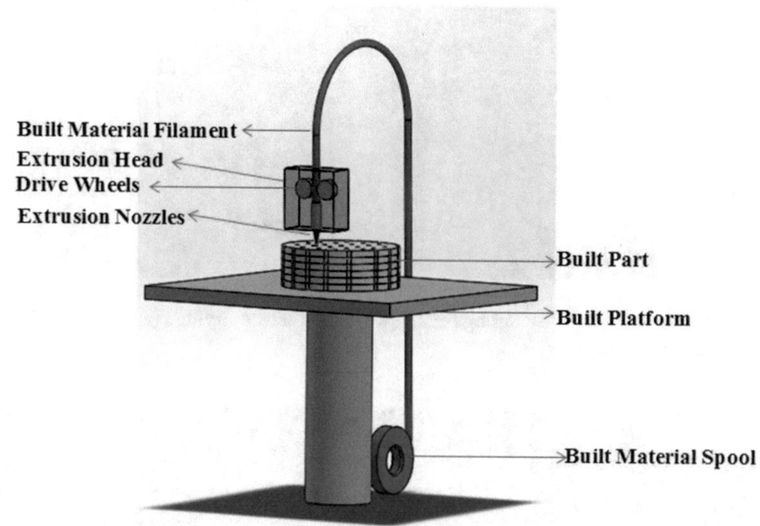

Fig. 9.13 Extrusion-based process (Liu et al. 2018)

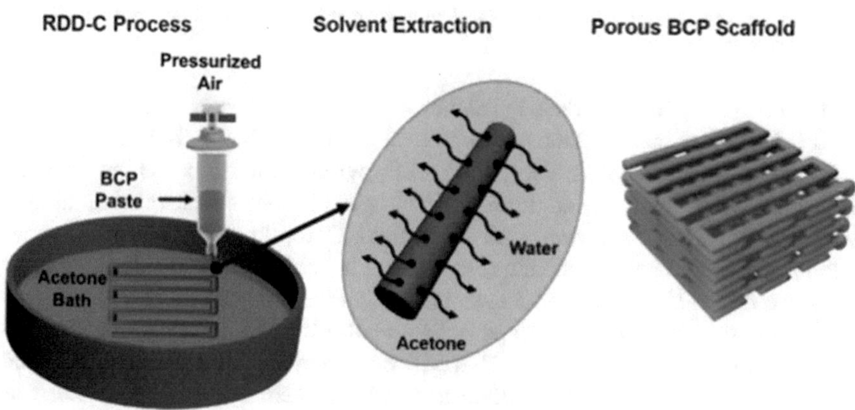

Fig. 9.14 Representation of material's deposition during an extrusion-based process (Liu et al. 2018)

Powder bed fusion processes have been used several times for the fabrication of bioceramic scaffolds. Suwanprateeb et al. produced HA–apatite–wollastonite glass-ceramic scaffolds, where HA and A–W glass powder were mixed with maltodextrin as a binder. Then, the solution was dried and was grinded with a diameter of 70 μm [34, 70]. Finally, the solution was sintered with a layer thickness of 100 μm and bending strength of 50 to 150 MPa (Fig. 9.18). The biological tests showed that the produced scaffolds were nontoxic for the osteoblast cells [34, 70] (Fig. 9.19).

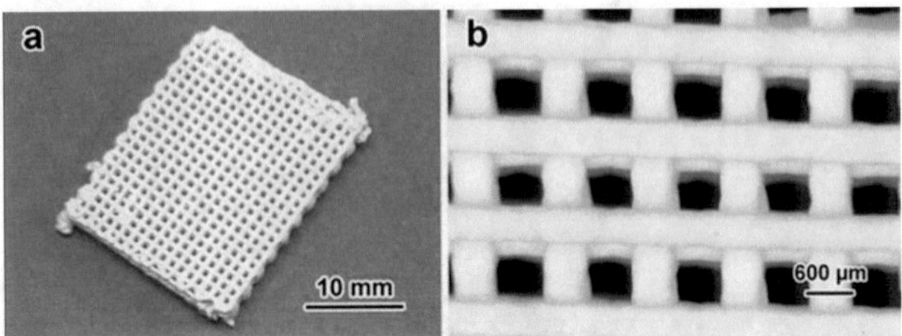

Fig. 9.15 Fabricated scaffolds through extrusion-based freeze drying. (**a**) Sample's view. (**b**) Sample's SEM image [27]

Fig. 9.16 (**a**, **b**) After deposition SEM images of fabricated scaffolds. (**c**) SEM image of sintered scaffold showing the dense of glass struts [27]

Bergmann et al. produced calcium phosphate–Bioglass 45S5 scaffolds (Fig. 9.20). The mixed solution was combined from 40 wt% of β-TCP to 60 wt% of 45S5 Bioglass, with orthophosphoric and pyrophosphoric acids as binders. Dicalcium pyrophosphate and dicalcium hydrogen phosphate were created through a cementing reaction and were sintered at 1000 ˚C, resulting in 3D scaffolds containing wollastonite and rhenanite (Fig. 9.21). The fabricated scaffolds had a layer thickness of 50–75 μm and bending strength of 14.9 ± 3.6 MPa [8, 34].

One more example of calcium phosphate scaffolds is HA–13–93 glass. The fabricated scaffolds had 0–60 wt% of HA, with a mixture of 6 wt% of polyacrylic and dextrin as binders, which were dried and milled. Then, 10 wt% of dextrin was added, and a solution of 7:1 water–glycerol was used for the printing, resulting in 150 μm of layer thickness [34, 76] (Fig. 9.22).

Furthermore, there are pure bioactive glass scaffolds (13–93 glass) fabricated with powder bed fusion process. The glass frit with the use of 6 wt% of dextrin was milled, dried, and grounded. The layers were connected with a solution of 7:1 water–glycerol and then sintered at 742 °C–795 ˚C, due to the fact that higher temperatures could cause crystallization effects [34].

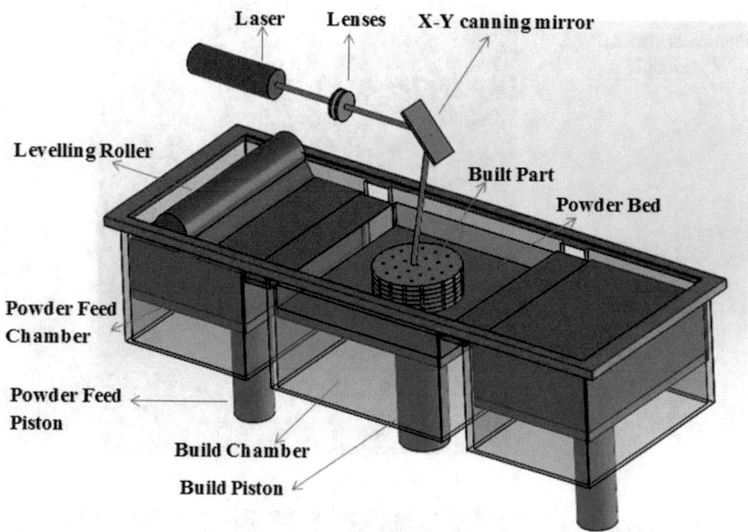

Fig. 9.17 Laser powder bed fusion process (LPDF) (Liu et al. 2018)

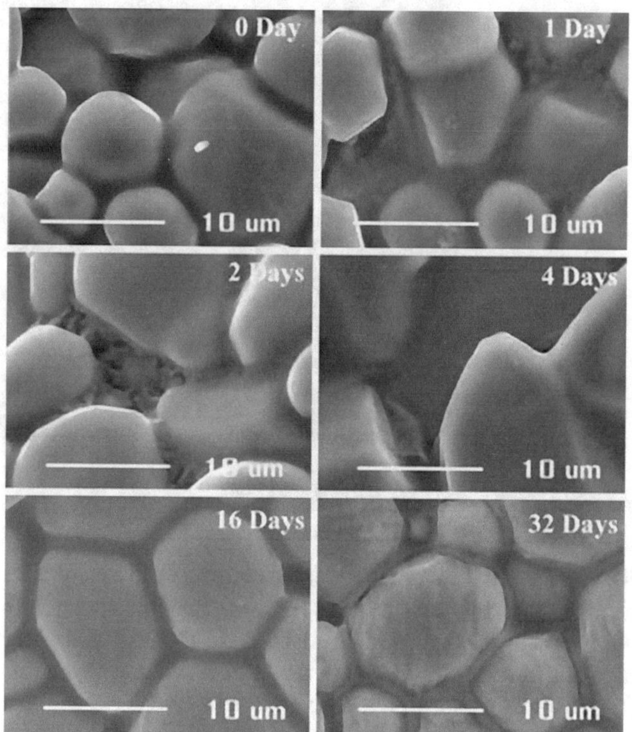

Fig. 9.18 Microstructure of 3D printed HA–A–W scaffolds soaking inside SBF [70]

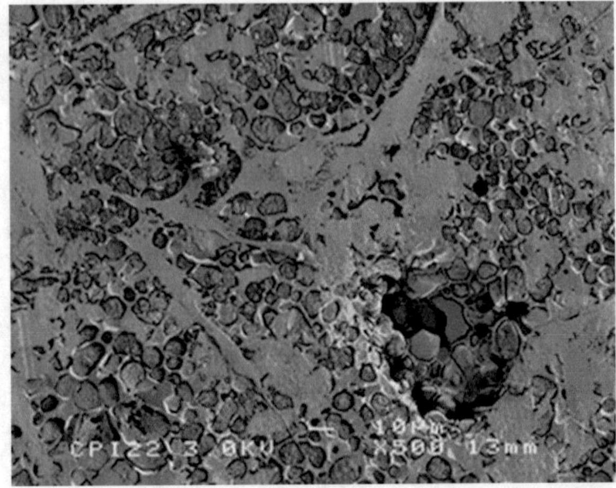

Fig. 9.19 SEM image of osteoblast cells attached on HA–A–W scaffolds surface [70]

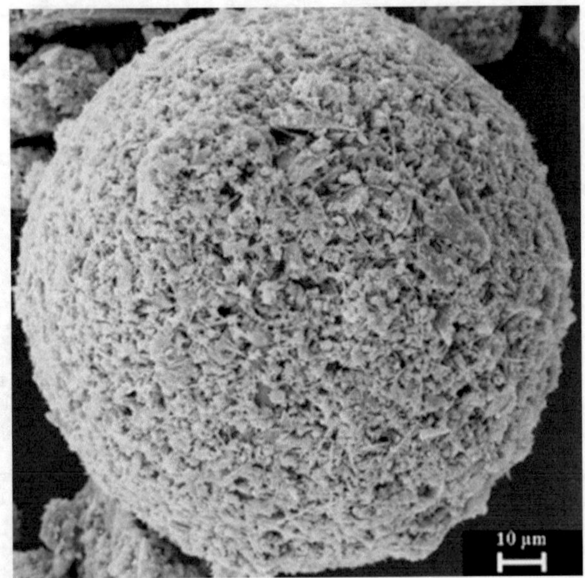

Fig. 9.20 SEM image of a bioglass–β-TCP granule [8]

9.8.3 Vat Photopolymerization Process

Vat photopolymerization is an additive manufacturing technique that has a promising future for the fabrication of scaffolds for medical applications [18]. Vat photopolymerization process technique due to the selective photopolymerization can fabricate large designed scaffolds and parts on microscale with millimeter details. One of its applications is an orthodontic treatment where custom removable and clean teeth aligners are fabricated for the patients. Moreover, through vat photopolymerization, the permeability, porosity, shape, pore size, and interconnectivity are controlled. Furthermore, it can be used for the encouragement of vasculogenesis

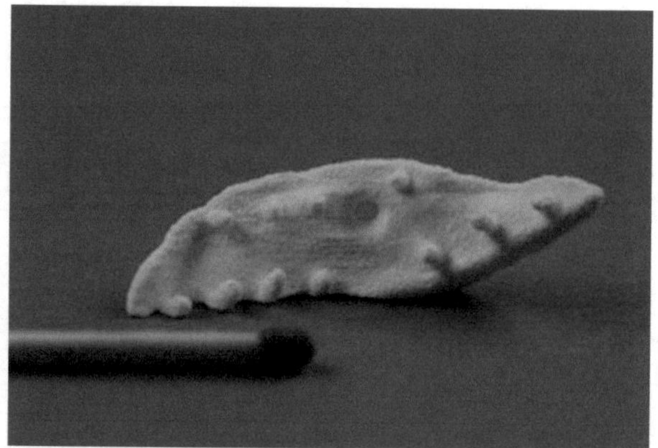

Fig. 9.21 Printed structure of a granulated bioglass–β-TCP granule [8]

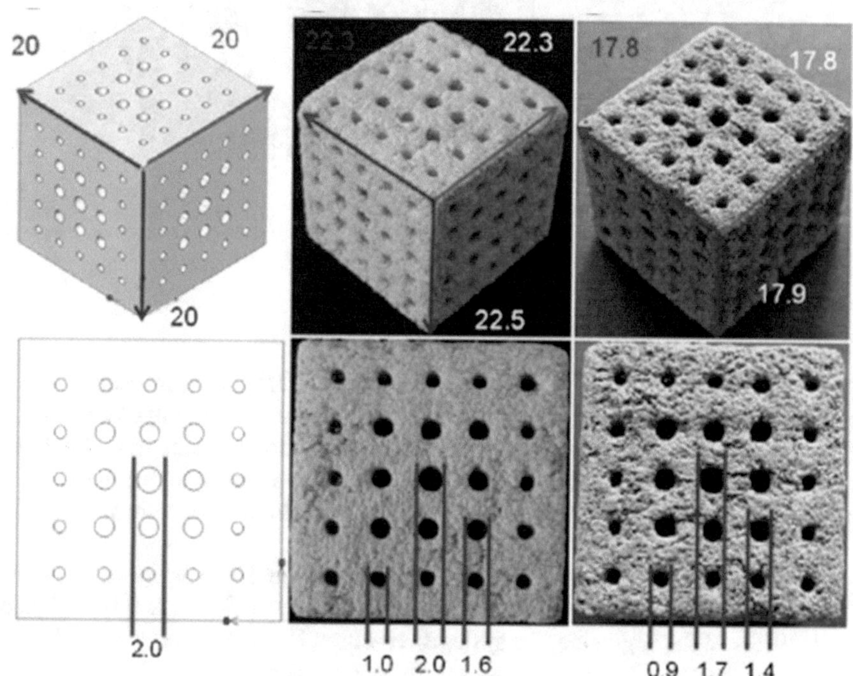

Fig. 9.22 CAD model, 3D-printed green body and sintered HA–glass structure [76]

and angio through the fabrication of vascular networks [18]. Also, through vat photopolymerization, the fabrication of aid device shells for hearing can be achieved.

Vat photopolymerization is an additive manufacturing process where a light, where in the most cases is in UV spectrum, cross-links selectively and solidifies layer by layer a photopolymer resin ([18], Sirrine et al. 2019) (Fig. 9.23). There are three light-patterning techniques that are developed for vat photopolymerization. Thy are as follows:

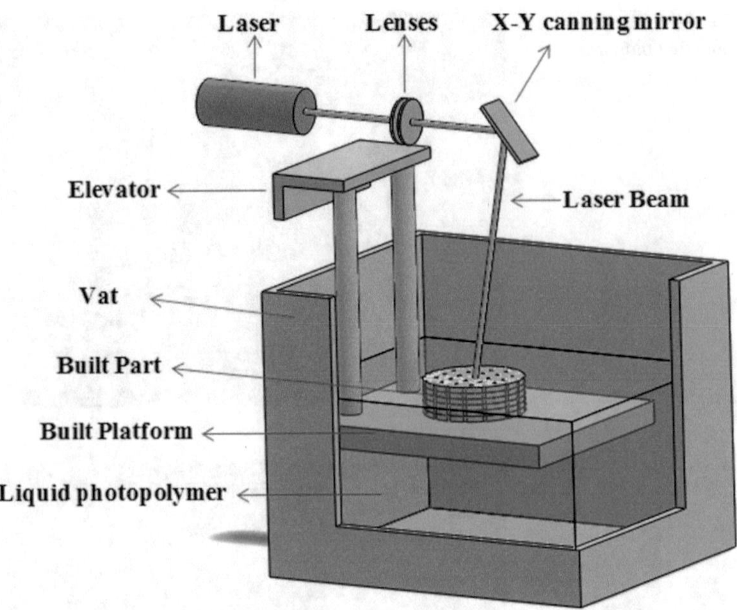

Fig. 9.23 Vat photopolymerization process (Liu et al. 2018)

- Vector scanning
- Mask projection
- Two photon

Vector scanning stereolithography is a technique that has been used in most machines, where a laser is scanned over the surface of a photopolymer resin vat ([18], Sirrine et al. 2019). A photopolymerization reaction happens when the laser has been scanned, and then the resin solidifies. When the layer has been fully scanned, the build stage is going deeper inside the photopolymer resin vat to build the next layer upon the previous layer (Fig. 9.24).

Mask projection stereolithography is a technique where the surface of the polymer vat irradiates simultaneously with the use of a dynamic mask ([18], Sirrine et al. 2019). At the beginning, the systems were used for their dynamic mask liquid crystal displays (LCD), while now the most machines use digital micromirror devices (DMD) and are used widely in projectors. DMD systems are micromirrors of large arrays that are able to rotate in both "off" and "on" positions ([18], Sirrine et al. 2019). In this case, an image of patterned light is produced by the reflection of the light that happens from several mirrors. Generally, mask projection stereolithography compared with vector scanning stereolithography is faster because of the simultaneous irradiation of the whole layer ([18]; Sirrine et al. 2019) (Fig. 9.24).

Thus, stereolithography uses single molecules for the absorption of one photon that leads to cross-linking and initiation, and there is also a reaction that includes the absorption of two photons of low energy, which is called two-photon stereolithog-

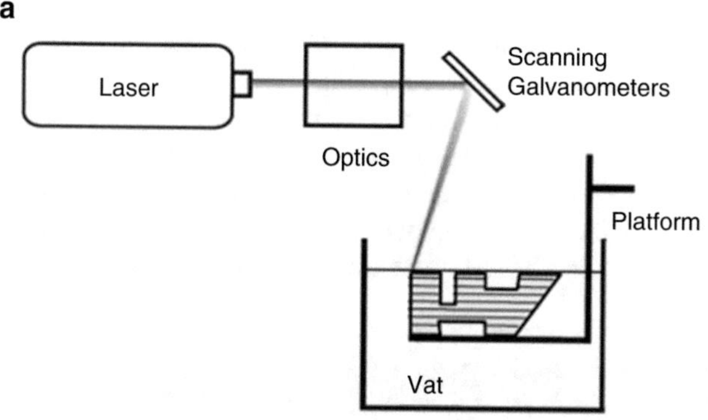

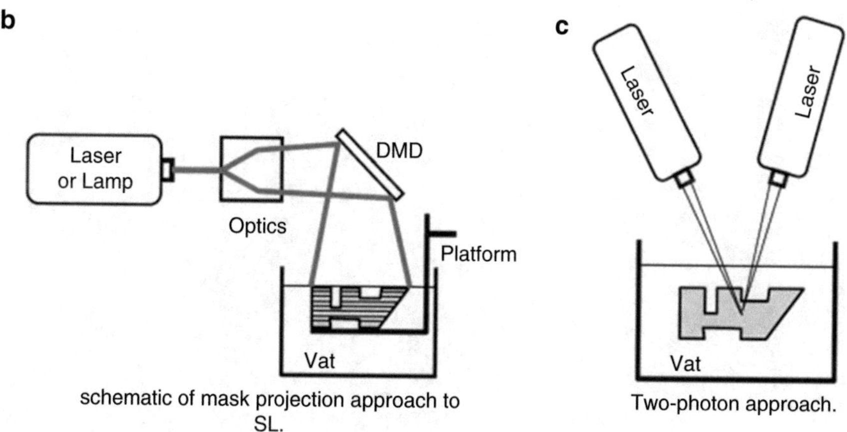

Fig. 9.24 (**a**) Vector scan stereolithography, (**b**) mask projection stereolithography, and (**c**) two-photon polymerization techniques [18]

raphy [18]. In two-photon stereolithography, a laser of ultrafast pulse is used for the creation of a high flow of photons in a small spatial and temporal volume for the absorption from the single molecule of the two photons. The two-photon absorption produced by a polymerization system of a two-photon at a wavelength of 780–820 nm has the same energy with the absorption of a single photon with the half wavelength. The fabricated parts of this technique are very small (>1mm^3) and are not used for the fabrication of scaffolds for implantation, but for the for understanding the interaction between the scaffold and the cells [18] (Fig. 9.24).

Thavornyutikarn et al., used Bioglass 45S5 (45 wt% SiO_2, 24.5 wt% Na_2O, 24.5 wt% CaO, and 6 wt% P_2O_5) mixed with photopolymer resin, as binder, and polyester-based copolymer, as dispersing, in which acidic groups were included [71]. The fabrication of the scaffolds was held on a stereolithography apparatus with the use of a laser UV source of 1.5 W. The produced scaffolds had various porous sizes between 435 μm and 870 μm, different designs (truncated octahedron, diamond-lie, rhombicuboctahedron, and cuboctahedron) (Figs. 9.25 and 9.26), porosity of 80%, and with compression strength of 2–12 MPa [71].

Moreover, Elomaa et al. with the use of stereolithography produced 3D scaffolds of photo cross-linkable residue of methacrylated polycaprolactone mixed with 0 wt%, 5 wt%, 10 wt%, and 20 wt% of bioactive glass, 0.10% of Orasol Orange G dye, and 3% of Lucirin TPO-L photoinitiator. The scaffolds were produced with the

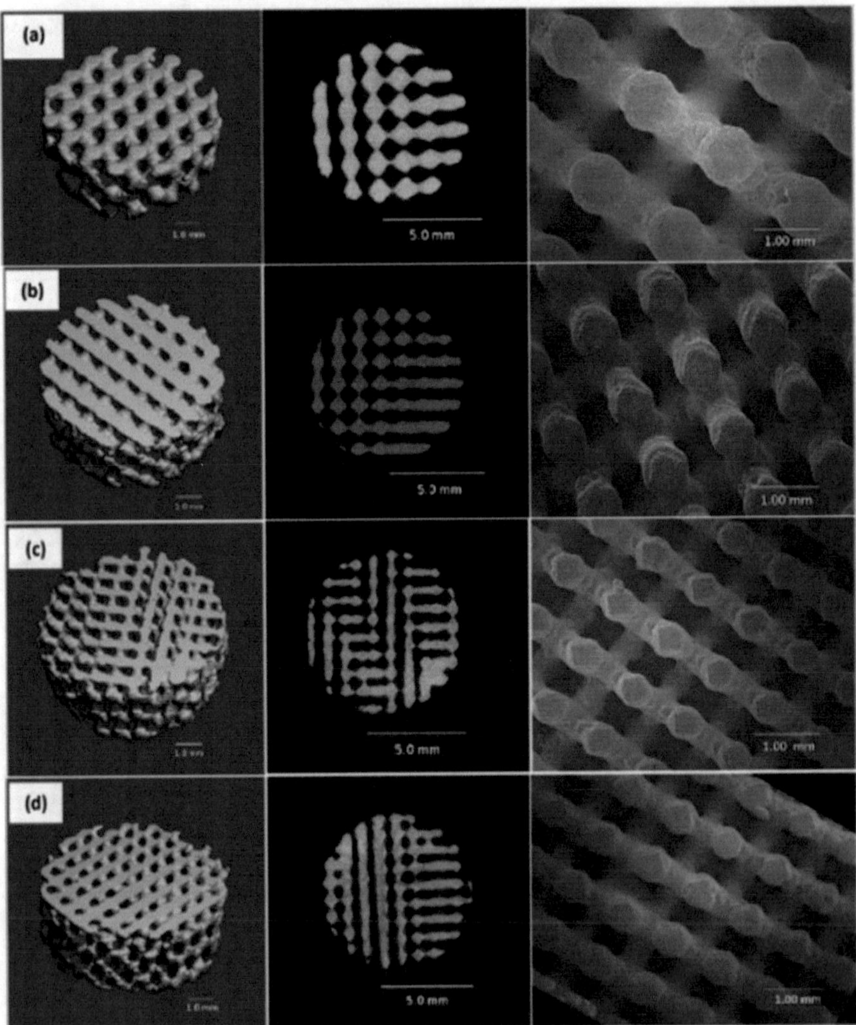

Fig. 9.25 CAD (left), 2D (center), and SEM (right) images of bioglass scaffolds with (**a**) 870 μm, (**b**) 700 μm, (**c**) 550 μm, and (**d**) 435 μm pore size [71]

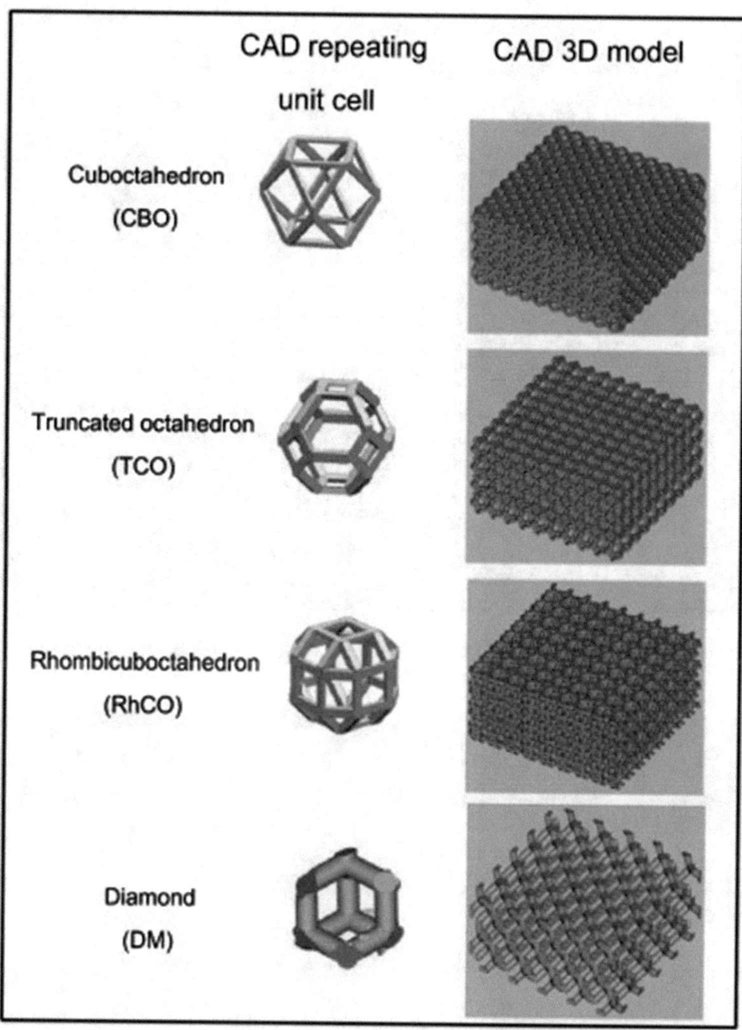

Fig. 9.26 Four geometrical structures of the fabricated scaffolds [71]

use of a local distance lens of 75 mm, where all the macromer particles were removed after the fabrication and the scaffolds were dried inside a vacuum [28] (Figs. 9.27 and 9.28). The produced scaffolds had 50 μm of layer thickness, mean pore size of 476–594 μm, porosity of 63–77%, and compression strength of 10.4 ± 0.8 MPa in dry form and 10.3 ± 10.6 MPa in wet form [28]. The biological tests showed an increase at the cell proliferation compared with the maximum rate of the samples (Fig. 9.29).

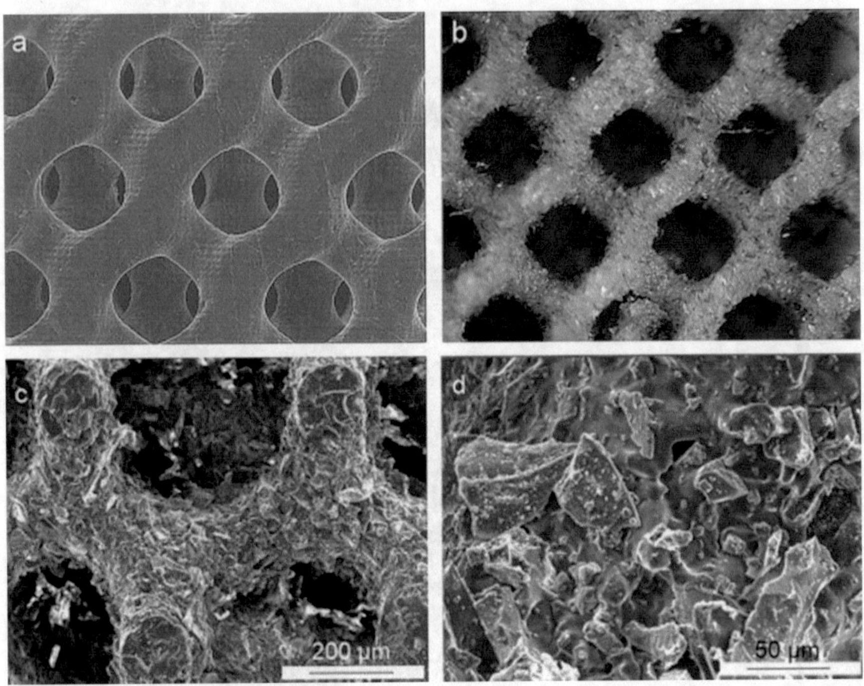

Fig. 9.27 (**a**) BG0 scaffold's SEM image, (**b**) BG20 scaffold's optical stereomicroscope image, (**c**) and (**d**) BG20 scaffold's SEM images [28]

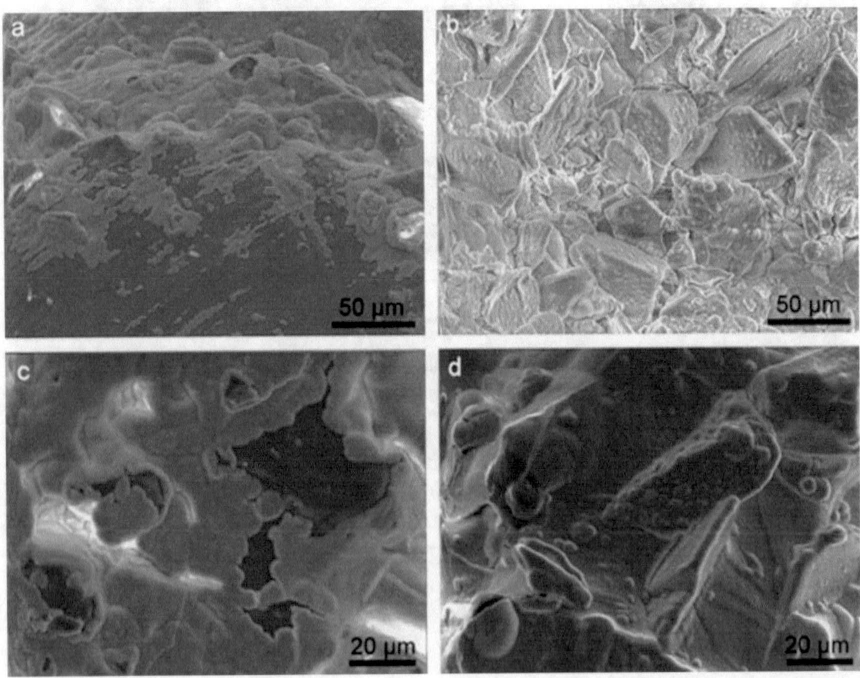

Fig. 9.28 SEM images of (**a**) 3 days BG20 scaffold, (**b**) 10 days BG20, (**c**) 21 days BG10, and (**d**) 21 days BG20 after their immersion inside SBF [28]

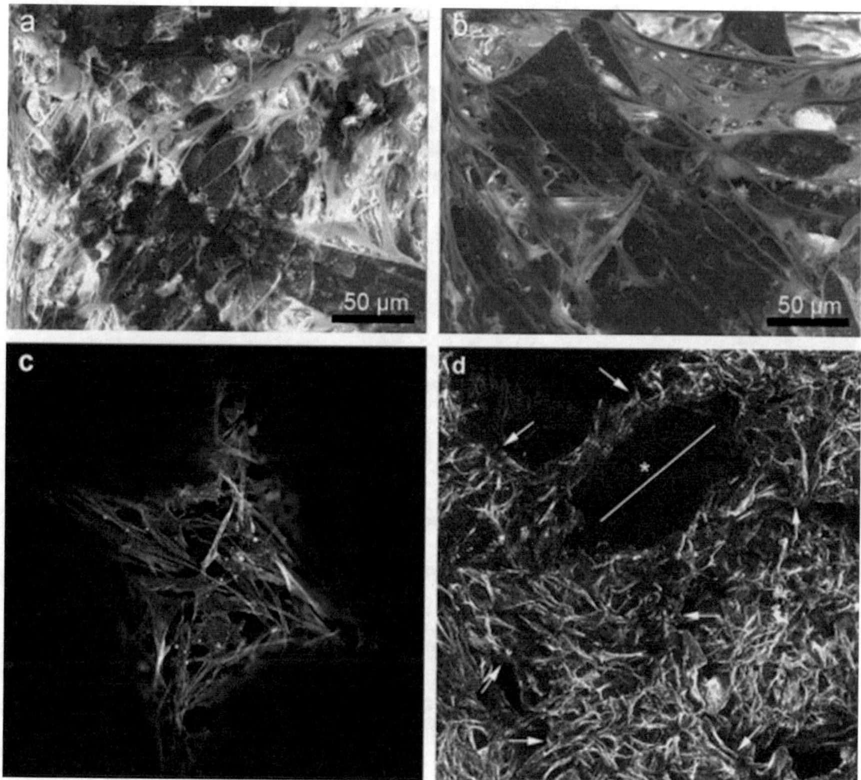

Fig. 9.29 SEM images 14 days cell culture of (**a**) BG10 scaffold, (**b**) BG20 scaffold, (**c**) BG10 scaffold's second layer, and (**d**) BG20 scaffold's middle layer [28]

9.9 Research Challenges

Despite of the exciting and amazing results of bioglass, there are some issues and challenges. One of the most important challenges is that the mechanical properties of bioglass scaffolds are very low and especially its compressive strength, making it unable for 100 wt% bioglass scaffolds. For that reason, bioglass has to be combined with other biocompatible and biodegradable polymers that can improve the mechanical properties of bioglass.

Furthermore, there is not a suitable protocol for the reduction of bioglass-derived glass-ceramic pH due to the fact that the toxicity of the scaffolds can lead to the death of the cells. Moreover, the sterilization techniques used can affect the chemical and physical abilities of bioglass scaffolds, although it is not in high extent.

One more challenge is that bioglass cannot be used with all the materials such as titanium due to the fact that the thermal expansion coefficient of the bioglass may not match with that of the metal leading to the separation of the two materials. Furthermore, bioglass during sintering process will crystallize, making it difficult for coating.

Finally, bioglass scaffolds have long degradation time, making it undesirable for bone tissue engineering applications. For that reason, its combination with slow degradation time polymers is necessary.

Acknowledgments This work has been supported by the Engineering and Physical Sciences Research Council (EPSRC) of the UK, the Global Challenges Research Fund (CRF), grant number EP/R01513/1.

References

1. M. Aebi, V. Arlet, J. Webb, *AOspine Manual: Principle and Techniques Volume 1* (Thieme, New York, 2007)
2. S. Ahn, Y. Kim, H. Lee, G. Kim, A new hybrid scaffold constructed of solid freeform-fabricated PCL struts and collagen struts for bone tissue regeneration: Fabrication, mechanical properties, and cellular activity. J. Mater. Chem. **22**(31), 15901 (2012)
3. T. Albrektsson, C. Johansson, Osteoinductive, osteoconductive and osseointegration. Eur. Spine J. **10**, 96–101 (2001)
4. D. Arcos, M. Vallet-Regí, Sol–gel silica-based biomaterials and bone tissue regeneration. Acta Biomater. **6**(8), 2874–2888 (2010)
5. A. Bailón-Plaza, M. van der Meulen, Beneficial effects of moderate, early loading and adverse effects of delayed or excessive loading on bone healing. J. Biomech. **36**(8), 1069–1077 (2003)
6. A. Bari, N. Bloise, S. Fiorilli, G. Novajra, M. Vallet-Regí, G. Bruni, A. Torres-Pardo, J. González-Calbet, L. Visai, C. Vitale-Brovarone, Copper-containing mesoporous bioactive glass nanoparticles as multifunctional agent for bone regeneration. Acta Biomater. **55**, 493–504 (2017)
7. P. Bartlett, *Bioelectrochemistry* (John Wiley & Sons, Chichester, 2008)
8. C. Bergmann, M. Lindner, W. Zhang, K. Koczur, A. Kirsten, R. Telle, H. Fischer, 3D printing of bone substitute implants using calcium phosphate and bioactive glasses. J. Eur. Ceram. Soc. **30**(12), 2563–2567 (2010)
9. S. Best, A. Porter, E. Thian, J. Huang, Bioceramics: Past, present and for the future. J. Eur. Ceram. Soc. **28**(7), 1319–1327 (2008)
10. J. Bico, U. Thiele, D. Quéré, Wetting of textured surfaces. Colloids Surf. A Physicochem. Eng. Asp. **206**(1–3), 41–46 (2002)
11. A. Boccaccini, J. Gough, *Tissue Engineering Using Ceramics and Polymers* (CRC Press, Boca Raton, 2007a)
12. A. Boccaccini, J. Gough, *Tissue Engineering Using Ceramics and Polymers* (Woodhead Publishing limited, Cambridge, UK, 2007b)
13. A. Boccaccini, X. Chatzistavrou, J. Blaker, S. Nazhat, Degradable and Bioactive Synthetic Composite Scaffolds for Bone Tissue Engineering, in *Degradation on Implant Materials*, (Springer, [Place of publication not identified], 2011)
14. S. Bose, S. Tarafder, Calcium phosphate ceramic systems in growth factor and drug delivery for bone tissue engineering: A review. Acta Biomater. **8**(4), 1401–1421 (2012)
15. D. Brown, R. Neumann, *Orthopedic Secrets* (Elsevier Health Sciences, London, 2004)
16. E. Bueno, J. Glowacki, Biologic foundations for skeletal tissue engineering. Synth. Lect. Tissue Eng. **3**(1), 1–220 (2011)
17. M. Cerruti, Surface characterization of silicate bioceramics. Philos. Trans. R. Soc. A Math. Phys. Eng. Sci. **370**(1963), 1281–1312 (2012)
18. N. Chartrain, C. Williams, A. Whittington, A review on fabricating tissue scaffolds using vat photopolymerization. Acta Biomater. **74**, 90–111 (2018)
19. W. Chen, *Oculoplastic Surgery* (Thieme, New York, 2001)

20. Q. Chen, I. Thompson, A. Boccaccini, 45S5 Bioglass®-derived glass–ceramic scaffolds for bone tissue engineering. Biomaterials **27**(11), 2414–2425 (2006)
21. P. Chu, X. Liu, *Biomaterials Fabrication and Processing Handbook* (CRC Press Taylor & Francis Group, Boca Raton [etc.], 2008a)
22. P. Chu, X. Liu, *Biomaterials Fabrication and Processing Handbook* (Taylor & Francis, Boca Raton, 2008b)
23. L. Claes, C. Heigele, C. Neidlinger-Wilke, D. Kaspar, W. Seidl, K. Margevicius, P. Augat, Effects of mechanical factors on the fracture healing process. Clin. Orthop. Relat. Res. **355S**, S132–S147 (1998)
24. A. Cormack, A. Tilocca, Structure and biological activity of. Glas. Ceram. **370**, 1271–1280 (2012)
25. S. Di Nunzio, C. Vitale Brovarone, S. Spriano, D. Milanese, E. Verné, V. Bergo, G. Maina, P. Spinelli, Silver containing bioactive glasses prepared by molten salt ion-exchange. J. Eur. Ceram. Soc. **24**(10–11), 2935–2942 (2004)
26. J. Dias, P. Granja, P. Bártolo, Advances in electrospun skin substitutes. Prog. Mater. Sci. **84**, 314–334 (2016)
27. N. Doiphode, T. Huang, M. Leu, M. Rahaman, D. Day, Freeze extrusion fabrication of 13–93 bioactive glass scaffolds for bone repair. J. Mater. Sci. Mater. Med. **22**(3), 515–523 (2011)
28. L. Elomaa, A. Kokkari, T. Närhi, J. Seppälä, Porous 3D modeled scaffolds of bioactive glass and photocrosslinkable poly(ε-caprolactone) by stereolithography. Compos. Sci. Technol. **74**, 99–106 (2013)
29. H. Elsayed, P. Rebesan, M. Crovace, E. Zanotto, P. Colombo, E. Bernardo, Biosilicate® scaffolds produced by 3D-printing and direct foaming using preceramic polymers. J. Am. Ceram. Soc. **102**(3), 1010–1020 (2018)
30. M. Favus, *Primer on the Metabolic Bone Diseases and Disorders or Mineral Metabolism* (American Society for Bone and Mineral Research, Washington, DC, 2003)
31. Q. Fu, E. Saiz, M. Rahaman, A. Tomsia, Bioactive glass scaffolds for bone tissue engineering:State of the art and future perspectives. Mater. Sci. Eng. C **31**(7), 1245–1256 (2011)
32. L. Gerhardt, A. Boccaccini, Bioactive glass and glass-ceramic scaffolds for bone tissue engineering. Materials **3**(7), 3867–3910 (2010)
33. L. Gibson, M. Ashby, B. Harley, *Cellular Materials in Nature and Medicine* (Cambridge University Press, Cambridge, 2010)
34. R. Gmeiner, G. Mitteramskogler, J. Stampfl, Stereolithographic ceramic manufacturing of high strength bioactive glass. Int. J. Appl. Ceram. Technol. **12**(1), 38–45 (2014)
35. L. González-Torres, M. Gómez-Benito, M. Doblaré, J. García-Aznar, Influence of the frequency of the external mechanical stimulus on bone healing: A computational study. Med. Eng. Phys. **32**(4), 363–371 (2010)
36. D. Griffon, *Evaluation of Osteoproductive Biomaterials: Allograft, Bone Inducing Agent, Bioactive Glass and Ceramics* (University of Helsinki, Helsinki, 2002)
37. S. Hattar, A. Asselin, D. Greenspan, M. Oboeuf, A. Berdal, J. Sautier, Potential of biomimetic surfaces to promote in vitro osteoblast-like cell differentiation. Biomaterials **26**(8), 839–848 (2005)
38. L. Hench, Bioactive Ceramics. Ann. N. Y. Acad. Sci. **523**(1 Bioceramics), 54–71 (1988)
39. L. Hench, Bioceramics: From concept to clinic. J. Am. Ceram. Soc. **74**(7), 1487–1510 (1991)
40. L. Hench, Third-generation biomedical materials. Science **295**(5557), 1014–1017 (2002)
41. L. Hench, The story of bioglass®. J. Mater. Sci. Mater. Med. **17**(11), 967–978 (2006)
42. L. Hench, Chronology of bioactive glass development and clinical applications. New J. Glass Ceram. **03**(02), 67–73 (2013)
43. J. Hollinger, *An Introduction to Biomaterials*, 2nd edn. (CRC/Taylor & Francis, Boca Raton, 2011)
44. A. Hoppe. Bioactive Glass Derived Scaffolds with Therapeutic Ion Releasing Capability for Bone Tissue Engineering, Thesis, 2014

45. T. Huang, M. Rahaman, N. Doiphode, M. Leu, B. Bal, D. Day, X. Liu, Porous and strong bioactive glass (13–93) scaffolds fabricated by freeze extrusion technique. Mater. Sci. Eng. C **31**(7), 1482–1489 (2011)
46. J. Jones, Review of bioactive glass: From Hench to hybrids. Acta Biomater. **9**(1), 4457–4486 (2013)
47. J. Jones, L. Ehrenfried, L. Hench, Optimising bioactive glass scaffolds for bone tissue engineering. Biomaterials **27**(7), 964–973 (2006a)
48. J. Jones, L. Ehrenfried, P. Saravanapavan, L. Hench, Controlling ion release from bioactive glass foam scaffolds with antibacterial properties. J. Mater. Sci. Mater. Med. **17**(11), 989–996 (2006b)
49. J. Jones, O. Tsigkou, E. Coates, M. Stevens, J. Polak, L. Hench, Extracellular matrix formation and mineralization on a phosphate-free porous bioactive glass scaffold using primary human osteoblast (HOB) cells. Biomaterials **28**(9), 1653–1663 (2007)
50. V. KARAGEORGIOU, D. KAPLAN, Porosity of 3D biomaterial scaffolds and osteogenesis. Biomaterials **26**(27), 5474–5491 (2005)
51. R. Lanza, R. Langer, J. Vacanti, *Methods of Tissue Engineering* (Academic Press, San Diego, 2011)
52. S. Lee, D. Henthorn, *Materials in Biology and Medicine* (CRC/Taylor & Francis, Boca Raton, 2012)
53. S. Lu, M. Hu, I. Gogotsi, *Ceramic Nanomaterials and Nanotechnology III* (John Wiley & Sons, Hoboken, 2012)
54. J. Maroothynaden, J. Hench, The effect of micro-gravity and bioactive surfaces on the mineralization of bone. JOM **8**(1), 79–80 (2001)
55. A. Martínez, I. Izquierdo-Barba, M. Vallet-Regí, Bioactivity of a CaO−SiO2Binary glasses system. Chem. Mater. **12**(10), 3080–3088 (2000)
56. R. Narayan, *Biomedical Materials* (Springer, New York, 2009)
57. D. Njobuenwu, E. Oboho, R. Gumus, Determination of contact angle from contact area of liquid droplet spreading on solid substrate. Leonardo Electron. J. Pract. Technol. **6**(10), 29–38 (2007)
58. I. Ochoa, J. Sanz-Herrera, J. García-Aznar, M. Doblaré, D. Yunos, A. Boccaccini, Permeability evaluation of 45S5 Bioglass®-based scaffolds for bone tissue engineering. J. Biomech. **42**(3), 257–260 (2009)
59. G. Owens, R. Singh, F. Foroutan, M. Alqaysi, C. Han, C. Mahapatra, H. Kim, J. Knowles, Sol–gel based materials for biomedical applications. Prog. Mater. Sci. **77**, 1–79 (2016)
60. N. Pallua, *Tissue Engineering* (Springer, Heidelberg, 2011)
61. N. Patel, S. Best, I. Gibson, S. Ke, K. Hing, W. Bonfield, Preparation and characterisation of hydroxyapatite and carbonate substituted hydroxyapatite granules. Key Eng. Mater. **192-195**, 7–10 (2000)
62. R. Pereira, P. Bártolo, Traditional therapies for skin wound healing. Adv. Wound Care **5**(5), 208–229 (2016)
63. J. Polak, S. Mantalaris, S. Harding, *Advances in Tissue Engineering* (Imperial College Press, London, 2008)
64. B. Ratner, A. Hoffman, F. Schoen, J. Lemons, *Biomaterials Science: An Introduction to Materials in Medicine* (Academic Press, Amsterdam, 2012)
65. P. Saravanapavan, L. Hench, Low-temperature synthesis, structure, and bioactivity of gel-derived glasses in the binary CaO-SiO2 system. J. Biomed. Mater. Res. **54**(4), 608–618 (2001)
66. J. SHEA, S. MILLER, Skeletal function and structure: Implications for tissue-targeted therapeutics. Adv. Drug Deliv. Rev. **57**(7), 945–957 (2005)
67. M. Singh, H. Haverinen, P. Dhagat, G. Jabbour, Inkjet Printing: Inkjet Printing-Process and Its Applications. Adv. Mater. **22**(6), 673–685 (2010)
68. S. Sohrabuddin, *Mechanism of Nanoparticle and Nanotube Induced Cell Death* (ProQuest, [Place of publication not identified], 2008)

69. D. Sommerfeldt, C. Rubin, Biology of bone and how it orchestrates the form and function of the skeleton. Eur. Spine J. **10**(0), S86–S95 (2001)
70. J. Suwanprateeb, R. Sanngam, W. Suvannapruk, T. Panyathanmaporn, Mechanical and in vitro performance of apatite–wollastonite glass ceramic reinforced hydroxyapatite composite fabricated by 3D-printing. J. Mater. Sci. Mater. Med. **20**(6), 1281–1289 (2009)
71. B. Thavornyutikarn, P. Tesavibul, K. Sitthiseripratip, N. Chatarapanich, B. Feltis, P. Wright, T. Turney, Porous 45S5 Bioglass®-based scaffolds using stereolithography: Effect of partial pre-sintering on structural and mechanical properties of scaffolds. Mater. Sci. Eng. C **75**, 1281–1288 (2017)
72. P. Tran, L. Sarin, R. Hurt, T. Webster, Opportunities for nanotechnology-enabled bioactive bone implants. J. Mater. Chem. **19**(18), 2653 (2009)
73. M. Vallet-Regí, Ceramics for medical applications. J. Chem. Soc. Dalton Trans. (2), 97–108 (2001)
74. M. Vallet-Regí, A. Salinas, D. Arcos, From the bioactive glasses to the star gels. J. Mater. Sci. Mater. Med. **17**(11), 1011–1017 (2006)
75. M. Vallet-Regi, M.M. Garcia, M. Colilla, Biomedical applications of mesoporous ceramics: Drug delivery. Smart Mater. Bone Tissue Eng. **3**(1), 231 (2012)
76. A. Winkel, R. Meszaros, S. Reinsch, R. Müller, N. Travitzky, T. Fey, P. Greil, L. Wondraczek, Sintering of 3D-printed glass/HAp composites. J. Am. Ceram. Soc. **95**(11), 3387–3393 (2012)
77. S. Wu, X. Liu, K. Yeung, C. Liu, X. Yang, Biomimetic porous scaffolds for bone tissue engineering. Mater. Sci. Eng. R. Rep. **80**, 1–36 (2014)
78. I. Xynos, A. Edgar, L. Buttery, L. Hench, J. Polak, Ionic products of bioactive glass dissolution increase proliferation of human osteoblasts and induce insulin-like growth factor II mRNA expression and protein synthesis. Biochem. Biophys. Res. Commun. **276**(2), 461–465 (2000a)
79. I. Xynos, M. Hukkanen, J. Batten, L. Buttery, L. Hench, J. Polak, Bioglass ®45S5 stimulates osteoblast turnover and enhances bone formation in vitro: Implications and applications for bone tissue engineering. Calcif. Tissue Int. **67**(4), 321–329 (2000b)
80. S. Yang, K. Leong, Z. Du, C. Chua, The Design of Scaffolds for use in tissue engineering. Part II. Rapid prototyping techniques. *Tissue Eng* **8**(1), 1–11 (2002)
81. Z. Zhou, L. Chen, Morphology expression proliferation of human osteoblasts on bioactive glass scaffold. Mater. Sci. Poland **26**(3), 506–516 (2008)
82. M. Zilberman, *Active Implants and Scaffolds for Tissue Regeneration* (Springer Berlin, Berlin, 2011)

Index

P. J. Bártolo, B. Bidanda (eds.), *Bio-Materials and Prototyping Applications in Medicine*, https://doi.org/10.1007/978-3-030-35876-1

Printed in the United States
by Baker & Taylor Publisher Services